# Lignin Renewable Materials - Chemistry, Trends, Technology and Applications

*Edited by Tri-Dung (TD) Ngo and Behzad Ahvazi*

Published in London, United Kingdom

Lignin Renewable Materials - Chemistry, Trends, Technology and Applications
http://dx.doi.org/10.5772/intechopen.1008581
Edited by Tri-Dung (TD) Ngo and Behzad Ahvazi

Contributors
Abdelaziz El-hoshoudy, Abiodun Olayinka Ogunsanya, Anny Leudjo Taka, Anuoluwapo Morenike Ajibola, Augustine Ben Okoubulu, Behzad Ahvazi, Chiosa Cletus Odili, Christopher W. J. Murnaghan, Deborah Osi-Goyeyi Daramola, Godwin Omoghegbe-Aigbojie Igbinigie, Jesubori Winner Sojobi, Jiawei Chen, Olasunkanmi Oriola Akinyemi, Oludayo Emmanuel Kayode, Olufemi Tajudeen Salami, Omotayo Omolara Ogunbiyi, Sunday Olayinka Oyedepo, Timothy Arome Ayegba, Toluwanimi Samuel Oyegunle, Tri Dung (TD) Ngo, Yu Chen



First published in London, United Kingdom, 2026 by IntechOpen
IntechOpen is the global imprint of INTECHOPEN LIMITED, registered in England and Wales, registration number: 11086078, 167-169 Great Portland Street, London, W1W 5PF, United Kingdom

For EU product safety concerns: IN TECH d.o.o., Prolaz Marije Krucifikse Kozulić 3, 51000 Rijeka, Croatia, info@intechopen.com or visit our website at intechopen.com.

British Library Cataloguing-in-Publication Data
A catalogue record for this book is available from the British Library

Lignin Renewable Materials - Chemistry, Trends, Technology and Applications
Edited by Tri-Dung (TD) Ngo and Behzad Ahvazi
p. cm.
Print ISBN 978-1-83635-472-7
Online ISBN 978-1-83635-471-0
eBook (PDF) ISBN 978-1-83635-473-4

# Meet the editors

Dr. Ngo holds a Bachelor's and Master's degree in Chemical Engineering from Ho Chi Minh City University of Technology and a Ph.D. in Mechanical Engineering from Concordia University, Montreal, Canada (2007). He is a Principal Research Scientist at InnoTech Alberta, a division of Alberta Innovates. With 28 years of experience in chemical and mechanical engineering, he serves as a project manager, technical lead, and advisor on numerous industrial, academic, and government-funded projects, both nationally and internationally. His research focuses on novel, high-performance, functionalized, and environmentally friendly materials for diverse applications. He has edited three books, contributed 16 book chapters, and authored or co-authored 115 publications in international journals and conference proceedings. He also holds several inventions and patents in materials science and engineering.

Dr. Behzad Ahvazi earned his B. Sc. (Honors) in Chemistry and Biochemistry with distinction from Concordia University and completed his Ph.D. in 1998 at the McGill Pulp and Paper Research Center, specializing in macromolecules and lignocellulosic materials. With more than 28 years of research and development experience, he has led multidisciplinary initiatives in biomass processing, green technologies, and sustainable biomaterials. His work has contributed to advancements in land reclamation, oil and gas, bio-composites, advanced materials, and next-generation building products. Dr. Ahvazi has held senior research positions in both government and private sectors and has served as an Adjunct Professor and research collaborator at the University of Alberta.

# Contents

# Preface

The growing global demand for sustainable materials and environmentally responsible technologies has heightened interest in lignin, one of the most abundant renewable aromatic biopolymers on Earth. As a major component of lignocellulosic biomass, lignin has traditionally been regarded as a low-value by-product of the pulp, paper, and biorefinery industries. However, recent scientific and technological advances have transformed this perspective, positioning lignin as a highly promising feedstock for the development of advanced functional materials, green chemicals, energy applications, and sustainable industrial solutions.

The book *Lignin Renewable Materials – Chemistry, Trends, Technology and Applications* aims to provide a comprehensive overview of current progress, challenges, and future opportunities in lignin research and industrial utilization. This volume brings together contributions from researchers and experts working across multidisciplinary fields, including chemistry, materials science, nanotechnology, environmental engineering, petroleum applications, modeling, and composite technologies. The chapters collectively highlight the evolving role of lignin in supporting the transition toward a circular bio-economy and sustainable development. The book consists of 6 chapters, each of which can be regarded as a self-contained unit. The chapters summarize recently developed research concerning Lignin renewable materials-chemistry, trends, technology, and applications. The book begins with foundational discussions of lignin as a renewable material, including its chemical structure, properties, and reactivity. These introductory chapters establish the scientific basis necessary for understanding the complexity and versatility of lignin systems. Subsequent chapters focus on emerging applications of lignin in advanced hybrid nanomaterials for water treatment, petroleum-related processes, sustainable composite fabrication, and predictive modeling approaches.

Chapter 1, "Introductory Chapter: Lignin-Renewable Materials", provides an overview of lignin, including its chemical structure, inherent complexity, and the challenges associated with its utilization. It also outlines research motivation, scope, and current trends in lignin-based materials.

Chapter 2, "Chemistry of Lignin", presents a comprehensive examination of the fundamental chemistry of lignin, focusing on its complex structure and diverse functional groups. It explores lignin biosynthesis in plants and highlights structural variations among softwood, hardwood, and grass lignins, including the three primary phenylpropane units: p-coumaryl, coniferyl, and sinapyl alcohols. The chapter also discusses key extraction methods such as kraft and sulfite pulping processes, as well as chemical modification techniques used to enhance lignin properties. In addition, advanced characterization techniques, including NMR and FTIR spectroscopy, are introduced to elucidate lignin's molecular composition.

Chapter 3, "Modelling Lignin: A New Approach to an Old Problem", provides an overview of key lignin bonding patterns that have been synthetically reproduced to support the development of catalytic systems aimed at lignin valorization into value-added products. The chapter highlights the diverse functional groups in lignin, which offer significant potential as feedstocks for further processing and chemical synthesis. It further explains that lignin's linkages and bonding patterns result from a largely random radical-mediated biosynthetic process, which contributes to its structural complexity. Various approaches to lignin depolymerization are discussed, with particular emphasis on the synthesis of representative linkages to generate more manageable fragments for catalytic studies and optimization of reaction conditions.

Chapter 4, "Advances in Lignin-Based Hybrid Nanomaterials as a Sustainable Approach for Water Treatment", examines lignin-based hybrid nanomaterials as sustainable and multifunctional materials with applications in energy, water treatment, biomedicine, and catalysis. The chapter reviews functionalization strategies for lignin and recent advances in the synthesis of lignin-based nanomaterials, including nanoparticles, nanocomposites, and polymer nanocomposites. These materials are explored as nano-adsorbents for the removal of organic pollutants (e.g., microplastics), inorganic contaminants (e.g., mercury ions), and microbial pathogens from water. Key challenges such as structural variability, adsorption efficiency, regeneration performance, and scalability are also discussed to guide future development of high-performance lignin-based materials for water purification.

Chapter 5, "Lignin as a Sustainable Filler in Composites: Fabrications, Mechanical Properties, and Applications", reviews recent advances in polymer composites reinforced with lignin, including their fabrication methods and applications. Lignin-based composites can be engineered with various morphologies and particle sizes, enabling their use in smart materials, biomedical devices, structural components, and energy storage systems. The growing interest in lignin as a reinforcing filler reflects the increasing emphasis on sustainable and environmentally friendly materials in advanced composite technologies.

Chapter 6, "Application of Lignin in Petroleum Processes", discusses the use of lignin in enhanced oil recovery (EOR) applications. Due to its amphiphilic structure, lignin can function as a surfactant, reducing interfacial tension between oil and water and improving oil displacement efficiency. Chemically modified lignin derivatives can enhance viscosity and improve mobility control during polymer flooding, thereby increasing sweep efficiency. In addition, lignin-based nanoparticles can modify rock wettability toward a more water-wet state, facilitating oil release. Owing to its biodegradability and low cost, lignin represents a promising alternative to conventional synthetic EOR agents. Recent studies on lignin-polymer composites and functionalized lignin systems under reservoir conditions further highlight its potential to enhance oil recovery while reducing environmental impact.

Together, these contributions demonstrate how lignin-based technologies are expanding beyond conventional applications into high-value industrial and environmental sectors.

A notable strength of this book lies in its interdisciplinary perspective. The integration of chemistry, engineering, materials science, and computational approaches reflects the modern direction of lignin research and emphasizes the importance of collaboration among academia, industry, and government institutions. The editors believe that this compilation will serve as a valuable resource for researchers, graduate students, industrial scientists, and policymakers interested in renewable materials and sustainable technologies.

We sincerely thank all contributing authors for sharing their expertise and scholarly efforts in preparing this volume. We also express our gratitude to the reviewers and editorial team whose support and dedication made this publication possible. We hope this book will stimulate further innovation, encourage collaborative research, and advance lignin science and its industrial applications worldwide.

**Tri Dung (TD) Ngo, Ph.D.**
Principal Research Scientist, Adjunct Professor,
InnoTech Alberta,
Faculty of Engineering,
University of Alberta,
Edmonton, Alberta, Canada

**Behzad Ahvazi, Ph.D.**
Biomaterials Development Scientist,
Chief Scientific Officer,
University of Alberta,
Edmonton, Alberta, Canada

Section 1

# Fundamentals of Lignin

Chapter 1

# Introductory Chapter: Lignin-Renewable Materials

*Tri Dung (TD) Ngo and Behzad Ahvazi*

## 1. Introduction

Lignin is one of the most abundant biopolymers on the Earth, second only to cellulose in natural prevalence. It constitutes approximately 15–30% of lignocellulosic biomass and plays a central role in plant physiology and ecology. Structurally, lignin is deposited primarily in the secondary cell walls of vascular plants, where it forms a complex, heterogeneous network that is intimately associated with cellulose and hemicellulose. Functionally, lignin serves as the binding matrix that confers mechanical strength, rigidity, and hydrophobicity to plant tissues, allowing them to withstand compressive forces, resist microbial degradation, and efficiently conduct water through vascular tissues such as xylem [1, 2]. The presence of lignin has been a decisive factor in the evolutionary success of terrestrial plants, enabling them to achieve upright growth and colonize diverse environments (**Figure 1**).

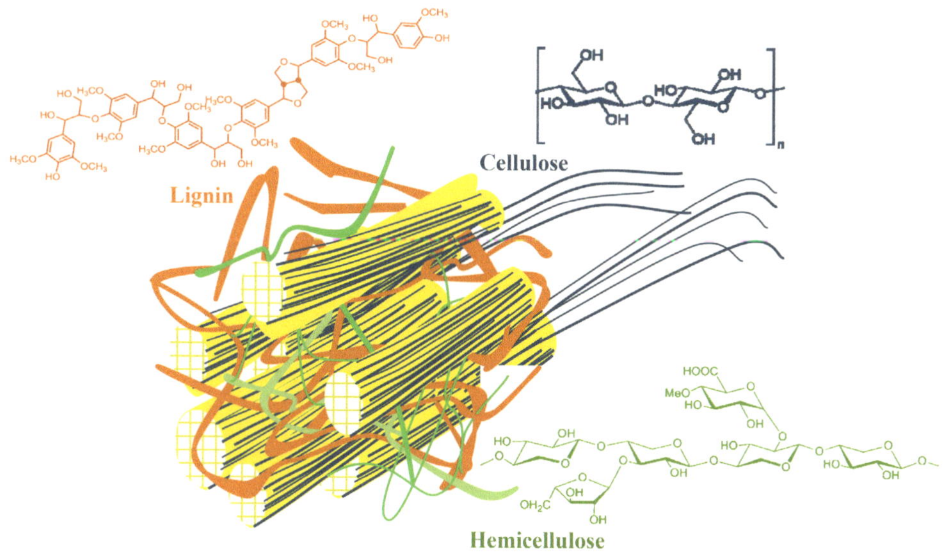


**Figure 1.**
*Composition of the plant cell wall.*

## 2. Chemical structure, complexity and challenges in lignin utilization

From a chemical standpoint, lignin is an amorphous, highly branched polyphenolic macromolecule of complex structure with high molecular weight. The chemical structure of lignin is highly irregular and proves extremely challenging. The lignin polymer consists primarily of phenylpropanoid monomers cross-linked together in three dimensions via a radical coupling process during its biosynthesis [3, 4]. Its backbone is derived from three primary monolignols: p-coumaryl alcohol, coniferyl alcohol, and sinapyl alcohol. These precursors are enzymatically polymerized via radical coupling reactions to yield p-hydroxyphenyl (H), guaiacyl (G), and syringyl (S) units, respectively. The resulting polymer network exhibits an extensive diversity of ether (β–O–4) and carbon–carbon (β–β, β–5, 5–5) linkages, making it structurally heterogeneous. This heterogeneity is influenced by species type (softwood, hardwood, grass), growth conditions, tissue type, and extraction method [5]. For example, softwood lignin is composed predominantly of guaiacyl units (80–90%), whereas hardwood lignin contains both guaiacyl and syringyl units, with syringyl units comprising up to 60% of the polymer. Grass and agricultural lignins are distinct in that they contain significant proportions of all three units (H, G, and S), in addition to ester-linked p-coumaric and ferulic acids that further complicate depolymerization (**Figure 2**).

Lignin is available in large quantities as a byproduct of the pulp and paper industries, with annual global production exceeding 70 million tons [6]. The most abundant industrial lignins are derived from kraft and sulfite pulping processes [7]. However, lignins obtained through these processes undergo significant structural modifications and are no longer identical to their native counterparts [8–10]. Due to its complex, heterogeneous, and partially undefined chemical structure, the industrial utilization of lignin has historically been limited [11–13].

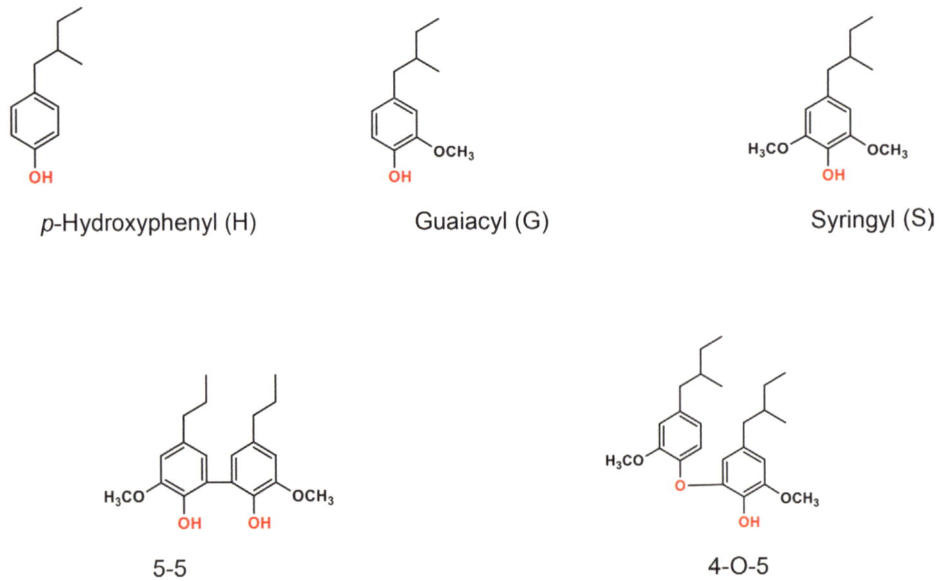


**Figure 2.**
*Structures of some lignin moieties.*

Currently, lignin generated as a pulping byproduct is utilized almost exclusively as fuel in recovery boilers to power evaporators and concentrate pulping liquors [13, 14]. While this practice is economically practical, it represents a low-value end use for a material with significant chemical potential. Approximately 98% of the lignin produced in pulp mills is burned on-site to generate process heat and recover chemicals, leaving only a small fraction available for material applications [15].

Recent years have witnessed a paradigm shift in how lignin is perceived within the scientific and industrial communities. No longer regarded merely as a waste stream or energy source, lignin is increasingly seen as a valuable raw material with broad potential for valorization. Driven by the imperatives of sustainability, circular economy initiatives, and the need for renewable alternatives to petroleum-derived aromatics, lignin research has gained remarkable momentum. Advances in lignin characterization, analytical chemistry, and depolymerization techniques are enabling a more detailed understanding of lignin structure-function relationships and providing routes to generate well-defined product streams from this complex biopolymer [16].

Valorization strategies for lignin can broadly be categorized into (i) depolymerization to produce low-molecular-weight aromatic monomers, (ii) functionalization and modification for use as macromolecular materials, and (iii) conversion into carbonaceous products such as activated carbons and carbon fibers. Depolymerization approaches, including catalytic hydrogenolysis, oxidative cleavage, reductive depolymerization, and biological conversion, are under active investigation to produce phenolic monomers, vanillin, phenol substitutes, and biohydrogen. Parallel research efforts are focused on using lignin as a macromolecular precursor for bio-based resins, adhesives, polyurethane polyols, hydrogels, and advanced carbon materials. Such applications have the potential to displace petrochemical-derived counterparts, thereby reducing greenhouse gas emissions and enhancing the economic viability of lignocellulosic biorefineries [5, 15].

## 3. Research motivation, scope and trend

It is imperative to note that the physical and chemical properties of lignin are strongly influenced by the wood species, the botanical origin, and—most critically—the extraction and recovery processes [17]. These factors significantly affect lignin's chemical structure [18], functional groups [19], and overall reactivity.

Different lignins contain distinct types and quantities of functional groups, which play a key role in determining their chemical behavior. During delignification, the lignin macromolecule is fragmented within the wood matrix, leading to its dissolution. This process triggers condensation reactions that cause substantial structural rearrangements. The mechanisms involved in these reactions are primarily governed by the nature of the carbon-oxygen (C-O) and carbon-carbon (C-C) bonds in the C9 units, as well as the presence of several interconnected arylpropane units [8] linked via ether and C-C bonds [20]. Although reactions involving the cleavage of C-C bonds occur, they are relatively minor compared to the extensive fragmentation that results from the rupture of ether linkages [21].

Lignin isolated under different processes exhibits considerable variation in its physical and chemical characteristics, even within the same wood species. Such variability, along with lignin's unique structural features, makes it a highly versatile candidate for substituting petroleum-based feedstocks in the production of industrial biomaterials and bio-composites [22, 23].

The incorporation of lignin into polymers for biomaterial applications is not governed solely by the reactivity of its chemical composition and functional groups. Other complex factors—such as solubility, molecular weight, rheology, dispersity, and morphology—also play a crucial role [22, 23]. For instance, key polymer properties such as glass transition temperature (Tg), modulus, and tensile strength are directly related to molecular weight. Consequently, the molecular weight distribution of different lignin fragments significantly impacts their performance during copolymerization [24].

In several studies, lignins extracted under different conditions were analyzed to determine their hydroxyl-containing functional groups (e.g., phenolic, aliphatic, and carboxylic acid groups) using $^{31}P$ NMR spectroscopy. Quantitative classification of these functional groups is essential for predicting their compatibility with various processes. Furthermore, the development of lignin-derived value-added products has been intensively investigated for the sustainable production of green chemicals and novel bio-based materials [25–28].

Despite these promising advances, several scientific and technological challenges must be addressed before lignin can realize its full potential as a feedstock for high-value applications [29]. Chief among these challenges is the intrinsic heterogeneity of lignin, which leads to batch-to-batch variability in its physicochemical properties. Additionally, conventional pulping processes often cause condensation reactions and chemical modifications that decrease the reactivity and solubility of lignin, thereby limiting its applicability in downstream transformations. Overcoming these barriers will require the development of more selective fractionation and extraction technologies, integrated catalytic processes capable of producing narrow product distributions, and robust analytical tools for comprehensive lignin characterization.

## Author details

Tri Dung (TD) Ngo[1*] and Behzad Ahvazi[2,3]

1 InnoTech Alberta, Edmonton, Alberta, Canada

2 Department of Mechanical Engineering Advanced Water Research Laboratory (AWRL), University of Alberta, Edmonton, AB, Canada

3 Department of Agricultural, Food and Nutritional Science Edmonton, University of Alberta, Edmonton, AB, Canada

*Address all correspondence to: tridung.ngo@innotechalberta.ca

## References

[1] Ali S, Rani A, Dar MA, Qaisrani MM, Noman M, Yoganathan K, et al. Recent advances in characterization and valorization of lignin and its value-added products: Challenges and future perspectives. Biomass. 2024;**4**:947-977. DOI: 10.3390/biomass4030053

[2] Ngo T-D, Ho T. Lignocellulosic biomasses: Sustainable platform for bio-based materials with polypropylene. In: Garcia P, editor. Polypropylene: Properties, Uses and Benefits. Hauppauge, NY: Nova Science Publishers; 2016. pp. 123-154

[3] Freudenberg K. The constitution and biosynthesis of lignin. In: Kleinszeller A, Springer GF, Wittmann HG, editors. Molecular Biology, Biochemistry, and Biophysics. Vol. 2. Berlin–Heidelberg: Springer-Verlag; 1968. pp. 47-122

[4] Wardrop AB. Occurrence and formation in plants. In: Sarkanen KV, Ludwig CH, editors. Lignins: Occurrence, Formation, Structure and Reactions. New York, NY: Wiley Interscience; 1971. pp. 19-41

[5] Cao Y, Chen SS, Zhang S, Ok YS, Matsagar BM, Wu KCW, et al. Advances in lignin valorization towards bio-based chemicals and fuels: Lignin biorefinery. Bioresource Technology. 2019;**291**:121878. DOI: 10.1016/j.biortech.2019.121878

[6] Chiang VL, Puumala RJ, Takeuchi H, Eckert RC. Comparison of softwood and hardwood Kraft pulping. Tappi Journal. 1988;**71**(9):173-176

[7] Sarkanen KV. Precursors and their polymerization. In: Sarkanen KV, Ludwig CH, editors. Lignins: Occurrence, Formation, Structure and Reactions. New York, NY: Wiley Interscience; 1971. pp. 95-163

[8] Sakakibara A. Chemistry of lignin. In: Hon DN-S, Shiraishi N, editors. Wood and Cellulose Chemistry. New York: Marcel Dekker Inc.; 1991. pp. 113-175

[9] Lai YZ, Sarkanen KV. Isolation and structural studies. In: Sarkanen KV, Ludwig CH, editors. Lignins: Occurrence, Formation, Structure and Reactions. New York: Wiley Interscience; 1971. pp. 165-240

[10] Chen C-L. Lignin: Occurrence in wood tissues, isolation, reactions and structure. In: Lewis M, Goldstein IS, editors. Wood Structure and Composition. New York: Marcel Dekker Inc.; 1991. pp. 183-261

[11] Adler E. Lignin chemistry: Past, present, and future. Journal of Wood Science and Technology. 1977;**11**:169-218

[12] Gellerstedt G, Lindfors E-L, Lapierre C, Monties B. Structural changes in lignin during Kraft cooking. Part 2. Characterization by acidolysis. Svensk Papperstidning. 1984;**87**:R61-R67

[13] Smook GA. Handbook for Pulp and Paper Technologists. Vancouver, BC: Angus Wilde Publications; 1982

[14] Thielemans W, Wool RP. Lignin esters for use in unsaturated thermosets: Lignin modification and solubility modeling. Biomacromolecules. 2005;**6**:1895-1905

[15] Ramzan H, Usman M, Nadeem F, Shahzaib M, Rahman MU, Singhania RR, et al. Depolymerization of lignin: Recent progress towards value-added chemicals and biohydrogen production. Bioresource

Technology. 2023;**386**:129492. DOI: 10.1016/j.biortech.2023.129492

[16] Ullah M, Liu P, Xie S, Sun S. Recent advancements and challenges in lignin valorization: Green routes towards sustainable bioproducts. Molecules. 2022;**27**(18):6055. DOI: 10.3390/molecules27186055

[17] Gross GG. Biosynthesis of lignin and related monomers. In: Loewus FA, Runeckles VC, editors. The Structure, Biosynthesis and Degradation of Wood, Rec. Adv. Phytochem. Vol. 11. New York, London: Plenum Press; 1977. pp. 141-184

[18] Gierer J. Chemistry of delignification. Part 1. General concept and reactions during pulping. Journal of Wood Science and Technology. 1985;**19**:289-312

[19] Adler E. Structural elements of lignin. Industrial and Engineering Chemistry. 1957;**49**(9):1377-1383

[20] Marton J. Reactions in alkaline pulping. In: Sarkanen KV, Ludwig CH, editors. Lignins: Occurrence, Formation, Structure and Reactions. New York: Wiley Interscience; 1971. pp. 639-694

[21] Gierer J. Chemical aspects of kraft pulping. Journal of Wood Science and Technology. 1980;**14**:241-266

[22] Feldman D. Lignin and its polyblends – A review. In: Hu TQ, editor. Chemical Modification, Properties, and Usage of Lignin. New York, NY: Kluwer Academic, Plenum Publishers; 2002. pp. 81-99

[23] Wang J, St. John Manley R, Feldman D. Synthetic polymer-lignin copolymers and blends. Progress in Polymer Science. 1992;**17**:611-646

[24] Yoshida H, Mörck R, Kringstad KP, Hatakeyama H. Kraft lignin in polyurethanes II. Effects of the molecular weight of the Kraft lignin on the properties of polyurethanes from Kraft lignin-polyether triol-polymeric MDI system. Journal of Applied Polymer Science. 1990;**40**:1819-1832

[25] Borges da Silva EA, Zabkova M, Araújo JD, Cateto CA, Barreiro MF, Belgacem MN, et al. An integrated process to produce vanillin and lignin-based polyurethanes from Kraft lignin. Chemical Engineering Research and Design. 2009;**87**:1276-1292

[26] Glasser WG, Barnett CA, Muller PC, Sarkanen KV. The chemistry of several novel bioconversion lignins. Journal of Agricultural and Food Chemistry. 1983;**31**(5):921-930

[27] Wise DL, editor. Organic Chemicals from Biomass. In: Biotechnology Series 4. Menlo Park, California: Benjamin/Cummings; 1983

[28] Argyropoulos DS, editor. Materials, Chemicals and Energy from Forest Biomass. Washington, DC: ACS Symposium Series 954; American Chemical Society; 2007

[29] Satheesh Kumar MN, Mohanty AK, Erickson L, Misra ML. Lignin and its applications with polymers. Journal of Biobased Materials and Bioenergy. 2009;**3**(1):1-24

Chapter 2

# Chemistry of Lignin

*Chiosa Cletus Odili, Anuoluwapo Morenike Ajibola, Jesubori Winner Sojobi, Omotayo Omolara Ogunbiyi, Abiodun Olayinka Ogunsanya, Augustine Ben Okoubulu, Godwin Omoghegbe-Aigbojie Igbinigie, Deborah Osi-Goyeyi Daramola, Oludayo Emmanuel Kayode, Toluwanimi Samuel Oyegunle, Timothy Arome Ayegba, Olufemi Tajudeen Salami, Olasunkanmi Oriola Akinyemi and Sunday Olayinka Oyedepo*

## Abstract

This chapter provides a comprehensive exploration of the fundamental chemistry of lignin, exploring its complex structure of diverse functional groups. It examines the biosynthesis of lignin in plants, highlighting variations among softwood, hardwood, and grass lignin, such as the three primary phenylpropane units—p-coumaroyl, coniferyl, and sinapyl. This chapter also covers key extraction methods, such as the kraft and sulfite pulping processes. It further covers the chemical modification techniques to enhance lignin properties. Advanced characterization tools such as NMR and FTIR are introduced to analyze lignin's molecular composition by providing a thorough understanding of lignin's chemical nature. This will lay the groundwork for its effective utilization in various applications.

**Keywords:** structure, functionalization, characterization polymerization, chemical structure, functional group

## 1. Introduction

The main structural component of plant secondary cell walls is lignin, a very complex phenolic polymer that coexists with cellulose and hemicellulose (**Figure 1**). According to Boerjan et al. [2], it is essential for plant growth and terrestrial adaptation because it provides mechanical stiffness, hydrophobicity, and protection against microbial degradation.

About 30% of the organic carbon in the biosphere is made up of lignin, which is also the second most prevalent biopolymer of lignocellulosic biomass, after cellulose [2]. It is covalently bound to the carbohydrate/cellulose structure and is frequently dispersed in combination with hemicellulose around the cellulose strands in primary

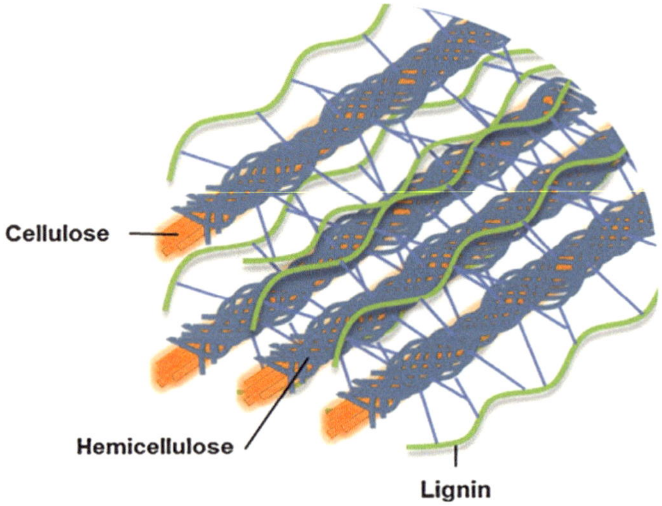


**Figure 1.**
*Lignin distribution in lignocellulosic biomass [1].*

and secondary cell walls. Lignin being an aromatic polymer is being noticed as a valuable precursor for bio-based products due to its plasticity in terms of structure, biodegradability, and renewable nature. Large volumes of lignin, a crucial component of plant biomass, are produced as a byproduct of the pulp and paper industry, offering a promising avenue for the development of sustainable materials. Lignin, being aromatic in nature and its relative abundance, can be used in making sustainable materials such as polyurethane foam, carbon fiber [3–5]. It may also be used in producing bio-based epoxy resin, lightweight composites [6, 7].

Furthermore, lignin nanoparticles (LNPs) are being developed for drug delivery, antioxidant packaging films, including hydrogels, owing to their phenolic content and bioactivity [8, 9].

In spite of these valuable properties, lignin is still underutilized, due to limited information about the chemistry of lignin. This research therefore tries to uncover the details, the chemistry, and various ways of improving its properties.

## 2. Structural features of lignin

### 2.1 Monolignols as building blocks

The formation of lignin occurs through the oxidative polymerization of three main monolignols (**Figure 2**) such as p-coumaryl, coniferyl, and sinapyl alcohols, which originate from the phenylpropanoid pathway [2, 11].

These monolignols vary in the number of methoxy groups attached to their aromatic rings, affecting the composition and properties of lignin across different plant species [12].

a. p-Coumaryl alcohol (H-Unit): Has a chemical name 4-hydroxycinnamyl alcohol, with a molecular formula $C_9H_{10}O_2$. It has a phenyl ring with a hydroxyl group at

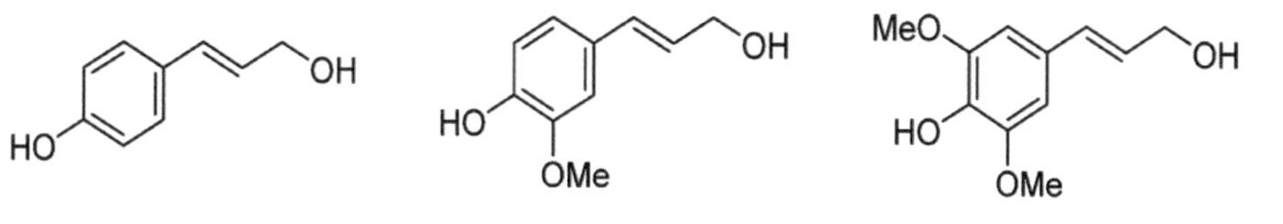


**Figure 2.**
*The chemical structures of the three primary monolignols found in lignin [10].*

the para position and a side chain (-CH=CH-$CH_2$OH) [2, 13]. It has the following key features: Lacks methoxy groups, forms H-lignin units upon polymerization. It is commonly found in grasses and other monocots, but generally in lower abundance compared to G and S units [14]. It provides basic structural framework, with less condensed and reactive compared to G and S units due to fewer functional groups [12].

b. Coniferyl alcohol (G-Unit): 4-hydroxy-3-methoxycinnamyl alcohol, with molecular formula $C_{10}H_{12}O_3$. It possesses one methoxy group at the aromatic ring's meta-position (3-position) [12]. Coniferyl alcohol is most abundant in gymnosperms (softwoods), which forms G-lignin units upon polymerization. It is capable of forming various linkages (e.g., β-O-4, β-5, β-β), contributing to the complexity of lignin structure [11]. It is highly reactive due to the presence of free ortho positions and contributes to cross-linking and rigidity in lignin polymer [13, 14].

c. Sinapyl alcohol (S-Unit): 4-hydroxy-3, 5-dimethoxycinnamyl alcohol, with a molecular formula $C_{11}H_{14}O_4$. It has two methoxy groups at the 3- and 5-positions of the aromatic ring [12]. Sinapyl alcohol is predominant in angiosperms (hardwoods). It forms S-lignin units, which are more linear due to restricted cross-coupling reactions [11].

## 2.2 Lignin polymerization and linkages

Lignin is formed through a radical-mediated dehydrogenative polymerization of monolignols (p-coumaryl, coniferyl, and sinapyl alcohols) within the plant cell wall [2]. The oxidative coupling of these monolignols generates a variety of inter-unit linkages, with the β-O-4 (beta-aryl ether) bond being the most prevalent and the main target for the degradation mechanism, comprising approximately 45–60% of linkages in native lignin [12]. Other significant bonds include β-5 (phenylcoumaran), β-β (resinol), 5-5 (biphenyl or dibenzodioxocin), and 4-O-5 (diaryl ether) linkages, though these are generally less abundant [14]. Some of these linkages and their relative abundance in different wood species are shown in **Table 1**.

The diversity of these linkages arises from the non-enzymatic nature of lignin polymerization, where radical coupling reactions occur in a largely uncontrolled fashion, influenced by substrate availability, redox conditions, and surrounding environment [14]. Lignin is composed of various inter-unit linkages that connect phenylpropanoid units, forming a highly branched network. The most common

| Linkage | Softwood | Hardwood | Grass |
|---|---|---|---|
| β-O-4 | 45–50% | 60–65% | ~45–55% |
| β-5 and 5-5 | 20–25% | ~10–15% | ~10–15% |
| β-β | ~5–10% | ~5–10% | ~5–10% |
| Hydroxycinnamate esters | Absent | Absent | High (pCA and FA) |

**Table 1.**
*Relative abundance of lignin linkages.*

lignin linkages (**Figure 3**) include β-O-4, 5-5, β-5, β-β, and 4-O-5 bonds, all of which contribute to the stability and reactivity of lignin.

## 2.3 β-O-4 linkage

The β-O-4 bond is the most prevalent linkage in lignin, comprising about 50–60% of all ether bonds. It connects two lignin monomers via an ether bond between the β-carbon of one unit and the 4-position of another aromatic ring. This linkage is weaker in comparison with C-C bonds, therefore making it a primary target for lignin depolymerization strategies [10].

## 2.4 5-5 linkage

This is a linkage that forms between two aromatic rings at their C6 positions. The bond is highly resistant to cleavage because of its stability, thereby making lignin degradation more difficult. The presence of 5-5 linkages contributes to the rigidity of lignin and mostly found in condensed lignin structures [16].

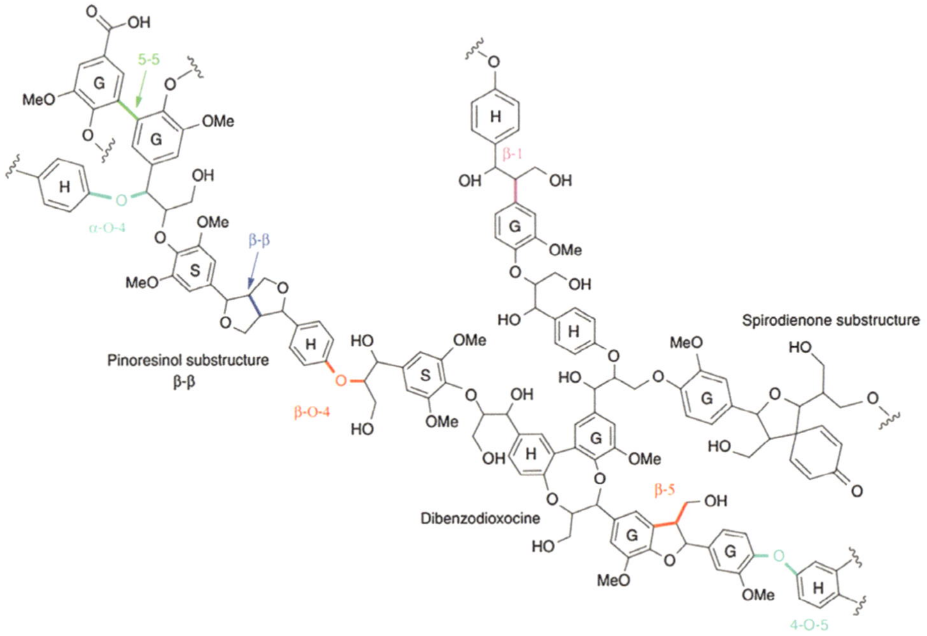


**Figure 3.**
*Lignin's main linkages [15].*

### 2.5 β-5 linkage

The β-5 bond consists of carbon-carbon linkage that connects β-carbon of one lignin unit to the fifth position of another aromatic ring. This bond is usually very small than β-O-4, but plays a major role in lignin's structural complexity. The β-5 linkage is usually associated with phenylcoumaran substructure, which affects lignin's reactivity during depolymerization [10].

### 2.6 β-β linkage

This linkage connects two lignin units at their β-carbons, forming resinol substructure. However, it is not very common but enhances lignin's resistance to degradation. Studies using density functional theory have revealed that β-β linkages undergo ring-opening reactions during thermal deconstruction [10].

### 2.7 4-O-5 linkage

This is an ether linkage that connects the 4-position of one aromatic ring to the 5-position of another. This bond is stronger as compared to β-0-4, due to radical formation on an aromatic carbon rather an aliphatic carbon. The bond dissociation energy of 4 = 9 = 5 linkage is higher than other bonds, therefore making them more resistant to cleavage [15].

A thorough understanding of lignin linkage is important for developing effective lignin depolymerization strategies. The β-0-4 bond is the most readily cleaved, while carbon-carbon linkages such as 5-5, β-0-5 and β-β show greater resistance to degradation. The 4-0-5 bond possesses enhanced stability as a result of its distinct ether structure. Recognizing these characteristics is essential for advancing effective lignin valorization strategies.

### 2.8 Heterogeneity and variability in structure

Lignin polymer shows considerable heterogeneity in its composition and structure across different species, cell types, and tissues [11]. This variation is due to differences in monolignol composition, proportions of various linkages, extent of cross-linking with other wall polymers, such as hemicellulose and proteins [17]. Additionally, lignin can incorporate phenolic molecule beyond the classic monolignols, such as ferulates, caffeates, and tricin, especially in grasses, which further enhances its structural diversity [18]. Lignin is an intricate, 3D, amorphous biopolymer majorly built from phenylpropanoid units. In contrast to cellulose and hemicellulose, it displays significant heterogeneity and variability in its chemical makeup. These characteristics present both challenges and potential advantages for lignin valorization.

#### *2.8.1 Variability in lignin between and within species*

Inter-Species Variability: This variability impacts depolymerization efficiency and product profiles (**Table 2**) [19].

#### *2.8.2 Intra-species and tissue-level variability*

Even within a single plant, lignin composition varies by tissue type (xylem, bark, fibers), cell wall layer (primary vs. secondary), and developmental stage.

| Plant Type | S/G/H ratio | Lignin content (%) |
|---|---|---|
| Hardwood | High S/G (~2:1) | 18–25 |
| Softwood | Almost all G | 25–35 |
| Grass | S:G:H ~ 1:1:1 | 15–30 |

**Table 2.**
*Different plant species and their lignin content.*

This is caused spatiotemporal regulation of monolignol biosynthesis and polymerization [20].

## 2.9 Technical variability: Effect of isolation methods

Isolation methods add further heterogeneity to technical lignin. Processes such as kraft pulping, organosolv, and soda treatments break native linkages and introduce new functionalities, such as phenolic OH, carbonyls, and sulfur groups, resulting in deviations from the native lignin structure.

## 2.10 Differences between softwood, hardwood, and grass lignins

Lignin, a complex phenolic polymer found in plant cell walls, exhibits considerable variation across different plant types (**Table 3**). Its composition, structure, and reactivity differ notably among softwoods (gymnosperms), hardwoods (angiosperms), and grasses (monocots). These distinctions arise from differences in monolignol precursors, inter-unit linkages, branching, cross-linking, and interactions with other cell wall components.

### *2.10.1 Softwoods (gymnosperms)*

Softwood lignin is composed almost exclusively of guaiacyl (G) units derived from coniferyl alcohol, leading to a more condensed, cross-linked polymer structure due to the availability of free ortho positions [14].

### *2.10.2 Hardwoods (angiosperms)*

Hardwood lignins consist of both guaiacyl (G) and syringyl (S) units, with the latter derived from sinapyl alcohol. The higher S-unit content leads to a less condensed and more linear polymer, due to limited cross-coupling at the aromatic ring [12].

| Plant type | S Units | G Units | H Units |
|---|---|---|---|
| Softwood | ~0% | 90–95% | 5–10% |
| Hardwood | 45–65% | 35–55% | <2% |
| Grasses | 25–50% | 25–50% | 5–35% |

**Table 3.**
*Variation of monolignols in different plants.*

### *2.10.3 Grasses (monocots)*

Grass lignin's are more complex and contain a mix of H, G, and S units, along with p-coumarates, ferulates, and other hydroxycinnamic acid esters. These non-canonical components introduce additional linkages and contribute to cross-linking with polysaccharides, making grass lignins chemically and structurally unique [21, 22].

## 3. Lignin biosynthesis: Pathways, polymerization, and transport mechanisms

Several interrelated routes, such as the tyrosine shortcut pathway and the phenylpropanoid pathway, are involved in the production of lignin. Polymerization and certain transport mechanisms are then employed.

### 3.1 The phenylpropanoid pathway

The process of lignin biosynthesis commences with the phenylpropanoid pathway (**Figure 4**), responsible for the synthesis of the three primary monolignols (**Figure 5**):

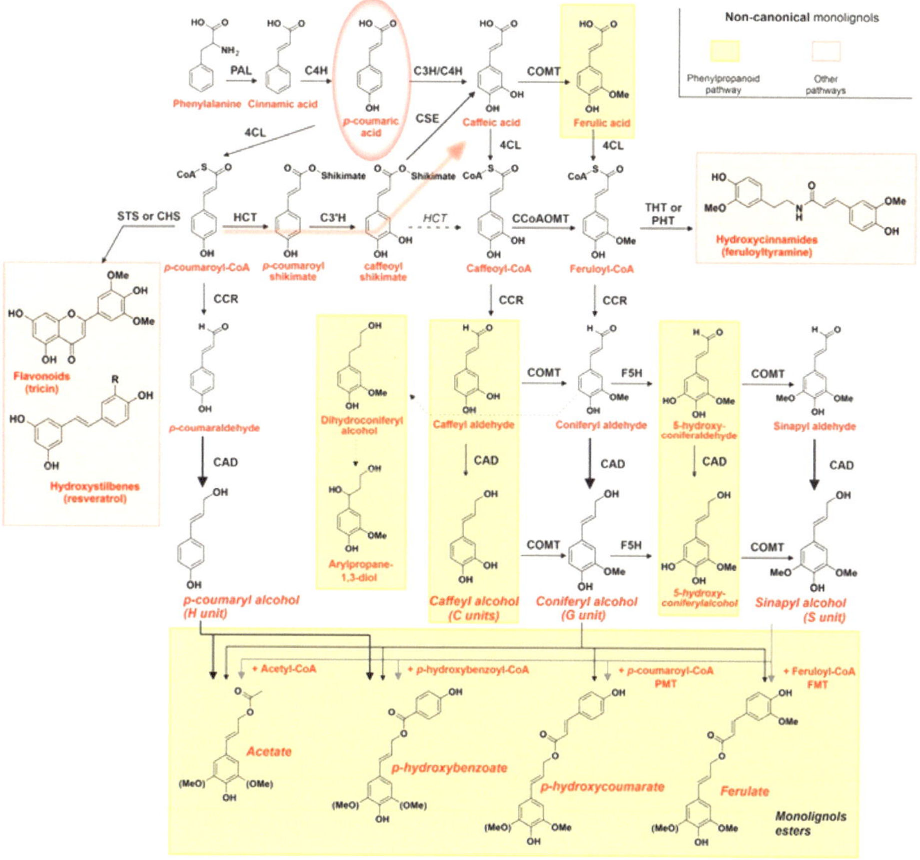


**Figure 4.**
*Phenylpropanoid pathway for lignin biosynthesis [23].*

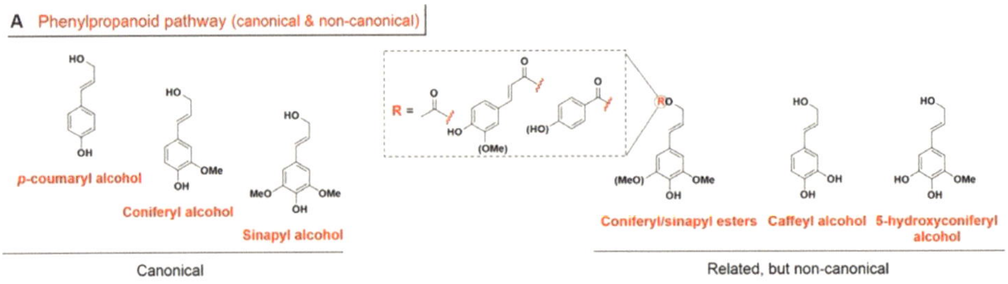


**Figure 5.**
*The primary monolignols [23].*

p-coumaryl alcohol (H units), coniferyl alcohol (G units), and sinapyl alcohol (S units) [21, 23]. The monolignols in question possess a phenolic core that is para-conjugated to a trans (E) double bond, which culminates in a hydroxymethylene group. Their distinction lies solely in the varying number of methoxy substituents [23].

The pathway begins with the conversion of phenylalanine (Phe) to cinnamic acid via phenylalanine ammonia-lyase (PAL), followed by hydroxylation to p-coumaric acid catalyzed by cinnamate 4-hydroxylase (C4H) [23]. Reduced lignin synthesis is the outcome of downregulating these enzymes [24]. Cinnamate derivatives are next reduced to aldehydes by cinnamoyl-CoA reductase, CCR), converted into CoA esters by 4-coumarate: CoA ligase, 4CL, and finally reduced to alcohols (cinnamyl alcohol dehydrogenase). It is noteworthy that plants lacking cinnamyl alcohol dehydrogenase (CAD) accumulate saturated non-canonical lignols, such as guaiacyl propane-1,3-diol and aldehydes [25–27].

Other enzymes including C3H, HCT, CSE, CCoAOMT, and COMT further modify intermediate downstreams of 4CL, ensuring proper lignin assembly [27]. It was also discovered recently that non-canonical phenolic chemicals that are integrated into lignin, for example, are hydroxystilbenes, hydroxycinnamic amides, and flavonoids [28]. The phenylpropanoid route is combined with polyketide or amino acid biosynthesis to produce these, which are derived from coumaroyl-CoA or feruloyl-CoA [29, 30].

## 3.2 The tyrosine shortcut pathway

The canonical process is circumvented by a more active metabolic pathway in grasses, such as *Brachypodium distachyon*. Bifunctional phenylalanine/tyrosine ammonia-lyase (PTAL) converts tyrosine straight to p-coumarate, obviating the necessity for C4H [29]. By diverting carbon flux away from rival metabolic pathways, such as benzenoid volatile production, this tyrosine-dependent pathway lowers energy consumption [27].

## 3.3 Lignin polymerization

Lignin polymerization occurs through radical coupling of monolignols, primary p-coumaryl, coniferyl, and sinapyl alcohols, yielding H, G, and S lignin units, respectively [27]. These activities may take place extracellularly or intracellularly, and is catalyzed by peroxidases and laccases [27]. H and G lignin are often found in vascular plant, and S lignin is found in angiosperms and certain lycophytes [31].

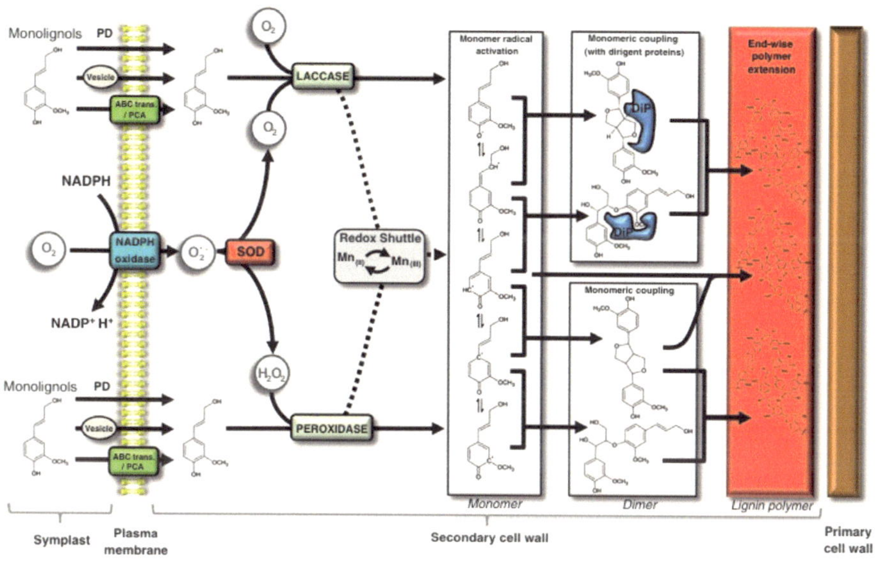


**Figure 6.**
*Lignin polymerization [33].*

Despite lacking lignin, bryophytes generate soluble phenylpropanoids, such as flavonoids [32].

### 3.4 Transport mechanism of monolignols

Monolignol synthesis occurs in the protoplast, but deposition is confined to the cell wall. Their transport mechanism has been proposed (**Figure 6**):

i. Passive diffusion (PD): Limited evidence supports this route, as monolignol portioning into liposomes is inefficient [32, 34].

ii. Vesicle-associated exocytosis: Suggested but not yet fully characterized

iii. Active transport
ABC Transporters: p-coumaryl alcohol is exported by ATP-dependent transporters, including ABC-G subtypes [35]. Proton-coupled antiporters: ABC transporter blockers do not reduce coniferin transport in xylem tissues, but proton gradient disruptors do [36].

## 4. Chemical properties of lignin

The creation of lignin-based biomaterials, the generation of biofuel, and the production of paper all depend on an understanding of the chemical characteristics of lignin, especially its solubility and degradation routes.

### 4.1 Solubility

In great part, the structural complexity and heterogeneity of lignin determine its solubility. Native lignin is typically insoluble in water due to its high molecular weight and extensive cross-linking with other cell wall components, such as cellulose and hemicellulose. However, solubility might differ greatly based on the chemical environment and the extraction technique. Generally, lignin shows poor solubility in neutral and acidic aqueous solutions but dissolves more readily in alkaline media. Alkaline environments increase lignin's solubility by breaking down its ether bonds. For instance, kraft lignin, which is obtained from the kraft pulping process, is soluble in aqueous sodium hydroxide due to the cleavage of β-O-4 linkages and introduction of hydrophilic functional groups [37].

Organic solvents such as dimethyl sulfoxide (DMSO), dioxane, and ethanol can also dissolve certain forms of lignin, especially when combined with heat or under specific pH conditions. Recently, ionic liquids and deep eutectic solvents have gained attention for their ability to dissolve lignin without extensive chemical modification, making them promising candidates for green chemistry applications [38].

### 4.2 Degradation pathways

Lignin degradation involves the cleavage of various inter-unit linkages, particularly ether and carbon-carbon bonds. β-O-4 (aryl ether), β-5 (phenylcoumaran), and β-β (resinol) structures are the most prevalent connections in lignin. Degradation can be accomplished by thermal, chemical, or biological methods.

### 4.3 Biological degradation

Lignin is primarily degraded in nature by white-rot and brown-rot fungi through enzymatic action. Enzymes such as lignin peroxidase (LiP), manganese peroxidase (MnP), and laccase play crucial roles in breaking down the complex lignin structure into smaller aromatic compounds. White-rot fungi are particularly effective, capable of mineralizing lignin into $CO_2$ and $H_2O$ [39].

### 4.4 Chemical degradation

Chemical methods involve the use of oxidizing agents such as hydrogen peroxide, ozone, or alkali treatments to cleave lignin linkages. Phenolic acids, aldehydes, and ketones are among the low-molecular-weight chemicals that are produced via oxidative depolymerization. In the paper and pulp industries, the kraft and sulfite processes are classic instances of chemical degradation.

### 4.5 Thermal degradation

Pyrolysis and hydrothermal liquefaction are thermal methods for lignin degradation. Lignin is thermally broken down by these mechanisms to produce char, gasses, and bio-oils. While hydrothermal liquefaction is carried out in subcritical water circumstances (200–374°C) under high pressure, pyrolysis usually takes place at temperatures above 400°C without oxygen [40].

## 5. Analytical techniques for lignin characterization

The structural heterogeneity and complex bonding nature of lignin demand a combination of analytical approaches to obtain a comprehensive understanding of its molecular architecture, functional groups, thermal stability, and reactivity. Since no single technique can capture the entirety of lignin chemistry, multiple techniques are essential.

Multiple analytical methods, such as spectroscopy methods, chromatographic techniques, thermal analysis, and wet chemical methods, have been involved in the investigation of lignocellulose composition and properties [41]. The principles of operation, application, advantages, limitations, and notable outcomes of each analytical technique are shown in **Tables 4-7** for spectroscopic, chromatography, thermal, and wet methods, respectively.

## 5.1 Spectroscopic methods

Spectroscopic techniques remain the cornerstone for lignin analysis, as they provide structural, functional, and compositional information at both molecular and supramolecular level.

### *5.1.1 UV spectroscopy*

UV spectroscopy is also called absorption spectroscopy in the ultra violet region in the range of 200–400 nm. Spectrometer measures the amount of light passing through the samples (I) and compares it with the amount of light before passing through the light (Io). These techniques usually measure samples in liquid form. Therefore, lignin is usually dissolved in a suitable solvent prior to the measurement. This method of measurement is usually simple; hence, it is commonly used to characterize lignin. Moreover, care must be taken in selecting appropriate solvent, as the spectrum stands a chance of being changed due to effect of solvent. Some common solvents usually used are water, dimethyl sulfides (DMSO), ethanol, and dichloromethane (DCM).

### *5.1.2 FTIR spectroscopy*

This is an absorption method, as a result of molecular vibration modes, which absorb particular frequencies of electromagnetic spectrum at different intensities. Several methods can be utilized to obtain FTIR data. Some of them are attenuated total reflectance (ATR) [50]. In this case, there is an intimate contact between the sample surface to be analyzed and a crystal. Another one is diffuse-reflectance infrared Fourier transform (DRIFT) spectroscopy. This method is well suited for characterizing powder. The powder is mixed with a purified salt and pressed to form a translucent pellet, in which a beam of infrared radiation is passed through. The two methods can be used to analyze lignin. It is very fast, needs little preparation, and small amount of sample is needed.

## 5.2 Chromatographic techniques

Chromatographic Techniques quantify lignin degradation product and evaluate molecular weight distributions.

### *5.2.1 Thermal analysis*

Differential scanning calorimetry is a thermos analytical method, in which the temperature of the sample being measured and the reference material is improved to

| S/N | Technique | Principle | Application to lignin | Advantages | Limitations | Notable outcomes | Reference |
|---|---|---|---|---|---|---|---|
| 1 | Nuclear magnetic resonance spectroscopy ($^{1}H$-NMR and $^{13}C$-NMR) | Magnetic resonance of certain atomic nuclei under a strong magnetic field and radiofrequency pulses. | Identifies aromatic rings, methoxy groups, and side chains of lignin. | Non-destructive; detailed structural data. | Requires solubilized lignin and expensive instruments. | Identification of S, G, H units and oxidation patterns. | [41] |
| 2 | Two-dimensional NMR (2D-NMR: HSQC, HMBC, TOCSY) | Correlates coupled nuclei through scalar or dipolar interactions. | Maps inter-unit linkages like β-O-4, β-β, and β-5. | High-resolution and detailed linkage data. | Complex interpretation and high cost. | Detailed structure elucidation across plant types. | [41] |
| 3 | Quantitative NMR (qNMR) | Quantifies signals relative to a standard for absolute functional group estimation. | Measures hydroxyl, methoxy groups and S/G/H ratios. | Accurate and reproducible without derivatization. | Requires calibration and high sample purity. | Quantification of hydroxyl groups for reactivity studies. | [42] |
| 4 | Fourier Transform Infrared Spectroscopy (FTIR) | Infrared absorption based on vibrational transitions of functional groups. | Identifies OH, C=O, and C-O-C bonds. | Fast and minimal sample preparation. | Overlapping peaks and semi-quantitative. | Used to distinguish between hardwood and softwood lignin. | [43] |
| 5 | Ultraviolet-visible spectroscopy (UV-Vis) | Measures absorbance of light in UV-Vis range by chromophores. | Analyzes conjugated aromatics and oxidation. | Simple, quick, and low cost. | Limited structural information. | Phenolic group estimation and lignin purity checks. | [44] |

**Table 4.**
*Spectroscopic methods.*

| S/N | Technique | Principle | Application to lignin | Advantages | Limitations | Notable outcomes | Reference |
|---|---|---|---|---|---|---|---|
| 1 | High-performance liquid chromatography (HPLC) | Separates analytes based on interactions with a stationary phase. | Quantifies monomers from lignin depolymerization. | High sensitivity and resolution. | Requires derivatization and pretreatment. | Determines S/G/H ratios and monomer yields. | [44] |
| 2 | Gel permeation chromatography (GPC) / size-exclusion chromatography (SEC) | Separates molecules based on hydrodynamic volume in solution. | Determines molecular weight distribution and polydispersity. | Effective for depolymerization/ purification studies. | Affected by solubility, standards, and aggregation. | Shows distinct molecular weight profiles in kraft vs. organosolv lignin. | [45] |

**Table 5.**
*Chromatographic techniques.*

| S/N | Technique | Principle | Application to lignin | Advantages | Limitations | Notable outcomes | Reference |
|---|---|---|---|---|---|---|---|
| 1 | Thermogravimetric analysis (TGA) | Monitors sample weight loss as temperature increases. | Analyzes decomposition behavior and char residue. | Good for comparing thermal stability. | Cannot identify decomposition products. | Shows lignin degrades at 200–300°C with high char yield. | [46] |
| 2 | Differential scanning calorimetry (DSC) | Measures heat flow differences during heating/cooling. | Determines glass transition and thermal transitions. | Useful in processing and compatibility studies. | Moisture and impurities affect results. | Tg range (100–180°C) varies across lignin types. | [47] |

**Table 6.**
*Thermal analysis techniques.*

| S/N | Technique | Principle | Application to lignin | Advantages | Limitations | Notable outcomes | Reference |
|---|---|---|---|---|---|---|---|
| 1 | Klason Lignin | Uses sulfuric acid to hydrolyze polysaccharides, leaving lignin residue. | Measures acid-insoluble lignin in biomass. | Widely used and reproducible. | Can overestimate due to pseudo-lignin or protein. | Standard method for lignin quantification in biomass studies. | [48] |
| 2 | Thioglycolysis | Breaks β-O-4 linkages under acidic thiolysis, yielding monomers. | Characterizes β-O-4 content and lignin units. | Reveals structural linkage patterns. | Requires GC/MS or HPLC for further analysis. | Used to assess structural integrity and monomer recovery potential. | [49] |

**Table 7.**
*Wet chemical methods.*

a constant rate. Thereafter, the rate of heat flow needed to increase the rate of sample and reference material to a constant rate is measured. It is used to measure the glass transition temperature (Tg) and change in heat capacity.

*5.2.2 Thermogravimetric analysis*

This method measures differences in weight of sample as a function of increasing temperature at a stable heating rate. It provides information, such as lignin degradation temperature and other phase transition temperatures.

## 6. Chemical modifications and reactions of lignin

Lignin's intrinsic heterogeneity, aromatic nature, and diverse functional groups allow a wide array of chemical modifications, enabling its transformation into value added materials and chemicals. These modifications fall broadly into depolymerization and functionalization, both of which are key to lignin valorization.

### 6.1 Depolymerization

Lignin depolymerization is a crucial strategy for transforming its complex macromolecular structure into smaller, value-added aromatic compounds [51]. It can also be referred to as the breakdown of lignin's complex, cross-linked structure into smaller aromatic compounds using chemical, biological, or thermal [51].

### 6.2 Hydrolysis

Depolymerization of lignin using hydrolysis is often done with water near supercritical conditions with a pressure of about 22 MPa [1]. Close to supercritical conditions, water serves as a good medium for fast and uniform reaction due to its properties, at these conditions like the ability to solubilize organic substance and little viscosity [15]. Different phenols such as guaiacol, catechol, and methoxy are produced using hydrolysis [52].

Hydrolysis typically targets ether bonds but is often limited due to the prevalence of stable C-C linkages in native lignin. Hydrolytic depolymerization, particularly acid-catalyzed hydrolysis, is generally less effective for lignin due to the presence of non-hydrolyzable ether and carbon-carbon (C-C) linkages. Acid- or base-catalyzed hydrolysis is more effective in modifying side chains or removing carbohydrate residues than breaking lignin's backbone [53]. However, it can serve to eliminate carbohydrate impurities or modify side chains [53]. Lignin, a naturally occurring polymer in plant cell walls, consists primarily of phenylpropane units linked by various ether and carbon-carbon (C-C) bonds.

### 6.3 Mechanism of hydrolysis in lignin depolymerization

Hydrolysis of lignin primarily focuses on breaking ether bonds, with β-O-4 linkages being the most prevalent target. This process can be facilitated or catalyzed by acids, bases, or enzymes, resulting in the production of phenolic monomers, oligomers, and various aromatic compounds [54].

i. Acid-Catalyzed Hydrolysis

ii. This involves strong acids such as sulfuric acid or hydrochloric acid, leads to cleavage of ether bonds and partial oxidation of lignin fragments, and generates phenolic compounds, aldehydes, and organic acids [55].

iii. Base-Catalyzed Hydrolysis
It uses alkaline solutions such as sodium hydroxide or ammonia, facilitates the breakdown of lignin into phenolic monomers and oligomers, and is often used in industrial processes for lignin valorization [55].

iv. Enzymatic hydrolysis
This involves lignin-degrading enzymes, such as laccases and peroxidase, which provide a selective and environmentally friendly approach. It produces bioactive compounds, with little side reactions [56].

### 6.4 Oxidation

Oxidative depolymerization, facilitated by oxidants such as $O_2$, $H_2O_2$, or polyoxometalates, produces aromatic aldehydes, ketones, and acids such as vanillin and syringaldehyde [56]. It uses hydrogenation usually metals, such as Ru/C or Ni/C, to cleave β-0-4 linkage and yield monomeric phenols, while preserving the aromatic rings. This route is promising for production fuels and polymer precursors [57]. Among the primary approaches for lignin depolymerization, oxidation plays a key role by breaking lignin into smaller molecular fragments using oxidative reactions.

#### *6.4.1 Mechanism of oxidative depolymerization*

Oxidative depolymerization primarily targets the cleavage of β-4 linkages, which are the most abundant in lignin. The process can be catalyzed using chemical oxidants, enzymes, or electrochemical methods, leading to the formation of phenolic monomers, carboxylic acids, aldehydes, and other aromatic compounds. The mechanism may be catalytic, enzymatic, or electrochemical oxidation.

### 6.5 Reduction

Reductive depolymerization, often facilitated by hydrogenation or catalytic transfer hydrogenolysis, targets ether bond (particularly β-0-4), resulting in monomeric phenols with preserved aromaticity [57].

#### *6.5.1 Mechanism of reductive depolymerization*

Reductive depolymerization primarily targets the cleavage of β-0-4 linkages, which are the most abundant in lignin. The process can be catalyzed by metal catalysts, hydrogen donors, or electrochemical methods, leading to the formation of phenolic monomers, alcohols, and other aromatic compounds. The mechanism includes the following.

i. Catalytic Hydrogenation
This process involves metal-based catalysts such as palladium, nickel, and ruthenium, facilitates selective cleavage of ether bonds and reduction of hydroxyl groups, and usually generates phenolic alcohols and lignin-derived hydrocarbons [54].

ii. Electrochemical Reduction
This process uses electrochemical cells to induce reduction reactions and offers a sustainable method for lignin valorization and produces high-value aromatic compounds.

iii. Biochemical Reduction
This involves microbial or enzymatic pathways for lignin degradation, provides an environmentally friendly approach, and generates bioactive compounds with minimal side reactions [54].

## 6.6 Functionalization (sulfonation, alkylation)

Lignin functionalization enhances its solubility, reactivity, and compatibility with other materials, enabling its use in diverse applications [58]. Sulfonation introduces sulfonic acid groups (-$SO_3H$) into the lignin macromolecule, significantly improving its water solubility [59]. Sulfonated lignin's (especially lignosulfonates) are widely used in dispersants, oil recovery agents, concrete plasticizers, and animal feed binders [60]. Sulfonation usually adds sulfonic acid groups to lignin, which enhances water solubility and allows for its use in dispersants, adhesives, and dye binder [60].

### *6.6.1 Mechanism of lignin sulfonation*

Sulfonation of lignin can be achieved using several chemical treatments, such as sulfuric acid, sodium sulfite, and sulfomethylation. The process primarily targets hydroxyl and ether groups leading to the formation of lignosulfonates, which are widely used in industrial applications.

Alkylation involves attaching alkyl group to hydroxyl or aromatic sites in lignin. This modification improves hydrophobicity, thermal stability, and compatibility with polyolefins and synthetic resins, enhancing lignin's performance in plastics and composite materials [61].

### *6.6.2 Mechanism of lignin alkylation*

Alkylation of lignin can be achieved through various chemical treatments, including methylation, ethylation, and propylation. The process primarily targets hydroxyl and ether groups, leading to the formation of alkylated lignin derivatives, which are widely used in industrial applications.

## 6.7 Emerging and green modification techniques

It is important to develop an environmentally friendly technique, with a minimum impact on the environment, when producing a desirable product from lignin. Therefore, it is important to minimize energy consumption and use less toxic materials [62].

#### *6.7.1 Sub- and supercritical fluids*

The use of supercritical fluid method is found useful in depolymerization of lignin, as it reduces condensation reactions. Sub- and supercritical fluids are usually achieved at high temperature and pressure [63]. They exhibit characteristic properties, such as low viscosity, small dielectric constant, which tend to depolymerize lignin.

#### *6.7.2 Ionic liquids and deep eutectic solvents*

Ionic liquids (IL) are recently regarded as a green technology that reduces processing costs, avoids environmental damage, and depolymerizes lignin through waste cleavage by cleaving to β-O-4 bonds. IL exhibit many qualities that are ideal for a green medium, including low formulation cost, miscibility, low viscosity, chemical inertness, and non-volatile nature [64]. They facilitate both depolymerization and functionalization under mild eco-friendly conditions. Since ILs are long-range, non-covalent, and interaction-based liquid formulations, recent investigations have shown that they are a good solvent for solubilizing lignin and fractionating lignocellulose into its fundamental components. Their cations and anions interact to break down the bonds of lignin and maintain their liquid phase condition, which is normally below 100°C [65].

Deep eutectic solvents (DES) are among the latest approach used in lignin depolymerization [66]. The most common methods are electrochemical catalysis, microwave irradiated and typical heating methods.

## 7. Industrial applications and valorization

### 7.1 Lignin in pulp and paper industry (kraft, sulfite processes)

Kraft and sulfite pulping is a process of dissolving lignin from wood, which results in the formation of cellulosic fibers for further use in commodity products, such as packaging, printing papers, and tissue [67].

Due to the ability to handle a wide variety of wood species and produce strong paper fibers, kraft process dominates the pulp and paper industry. To degrade lignin at high temperature and pressure, an alkaline solution of sodium hydroxide (NaOH) and sodium sulfite ($Na_2S$) is used. This process allows penetration of chemicals into the wood chips. The resulting pulp is known for its high strength and versatility [68].

In the paper industry, sulfite pulping is a major chemical process in which the main focus is on the removal of lignin from wood to obtain cellulose fibers, essential for paper production. Sulfurous acid and its salts known as sulfites are being employed to break down lignin, a complex polymer that holds cellulose fibers together in plant cell walls [69]. Through the introduction of hydrophilic sulfonate groups and hydrolytic cleavage of ether groups, the lignin becomes water soluble [67].

### 7.2 Lignin-derived products (vanillin, carbon fibers, adhesives)

Vanillin (4-hydroxy-3-methoxybenzaldehyde) is the most abundant commercialized mono-aromatic compound obtained from lignin on a large scale. Vanillin is widely used as a flavoring and fragrance agent in food products and a fragrance agent in cosmetics as well as a starting material for the synthesis of several second-generation fine chemicals [69]. Lignin-based vanillin is produced by the oxidative depolymerization of lignosulfonates from the sulfite pulping process; oxygen is used as an oxidant in the process, air, or nitrobenzene under alkaline conditions to cleave lignin's β-O-4 linkages, yielding vanillin and other aromatic aldehydes [70]. Industrial lignin is used as adhesive in wood panels due to the presence of phenolic groups in abundance. Phenol can be replaced with lignin in phenol-formaldehyde (PF) resins in wood composite adhesives [71]. Such adhesives can be used in the production of plywood, particleboard, and other kinds of wood composites. The broad use of lignin as adhesives is affected by the limited solubility of lignin in organic solvents. However, the solubility of kraft lignin in organic solvents can be improved by partial depolymerization through base-catalyzed depolymerization in supercritical methanol [72].

Carbon fibers have gained more attention due to the increasing demand for low-cost materials and weight-to-performance ratio. However, the high cost and their complex manufacturing processes of the precursors for carbon fibers (such as polyacrylonitrile (PAN), pitch, rayon, and phenolic resins) are the major factors limiting the large-scale application [73]. Lignin-based carbon fibers with varying diameter can be produced depending on the method of processing, such as dry-spinning, melt processing, or electrospinning. Lignin-based carbon materials, such as fibers, mats, nanofibers, and mesoporous carbon, have been successfully developed by a number of research groups. The choice of chemical modification of lignin precursors and carbonization parameters will determine the microstructure and properties of the carbon fibers [74].

## 7.3 Lignin in biofuels and biorefineries

As a byproduct of biorefineries, lignin can be produced in large scale [51]. Cellulose and hemicellulose polysaccharides are isolated from the biomass, the polysaccharides are then hydrolyzed to their monomers (saccharification), and microbial catalysts are used for the conversion of the sugars to fuels (e.g., ethanol, butanol) or chemicals (e.g., L- or D-lactate, succinate). Thermo-chemical pretreatment is used in commercial biorefineries. In the presence of a catalyst, the biomass is subjected to high temperature and high pressure [72, 75].

## 7.4 Emerging applications (nanomaterials, drug delivery)

Hydrogels are attractive materials for tissue engineering due to the intrinsic structural and compositional resemblance to the extracellular matrix. Hydrogels are attractive nanomaterials for tissue engineering. They are three-dimensional (3D) networks of polymer chains formed by cross-linking hydrophilic homopolymers, copolymers, or macromers to form insoluble polymer matrices [74, 76]. Hydrogels are biocompatible which explains why they are highly attractive for biomedical applications. The aqueous medium in which they embedded can protect cells drugs, peptides, and proteins from degradation [77].

A thermo-responsive hydrogel has been prepared using lignin having different gelling behavior at different temperatures. This was achieved by the reaction between acetic acid lignin with N-isopropylacrylamide (NIPAAm) in the presence of N, N′-methylenebisacrylamide (MBAAm) as the cross-linker and $H_2O_2$ as initiator.

Differential scanning calorimetry (DSC) analysis demonstrated that the lower critical solution temperature of the hydrogels was approximately 31°C. The developed material contained pore sizes between 20 and 100 μm that were tunable by altering the lignin concentration [78].

## 8. Environmental and sustainability aspects of lignin

### 8.1 Carbon sequestration and climate mitigation

Lignin's recalcitrant structure makes it a critical player in long-term carbon storage, particularly in forest ecosystems and agricultural soils. Unlike cellulose and hemicellulose, which decompose rapidly, lignin persists in soil organic matter for decades, effectively sequestering carbon [79]. In boreal forests, lignin-rich coniferous trees contribute significantly to carbon sinks, while in agricultural systems, crop residues high in lignin (e.g., wheat straw, corn stover) enhance soil carbon retention [80]. Recent studies estimate that lignin-derived carbon accounts for up to 30% of stable soil organic carbon pools, making it vital for climate change mitigation strategies (IPCC, 2021). One of such strategies is carbon capture and storage systems.

### 8.2 Lignin and carbon capture: A natural solution to a global challenge

Climate change has been a central topic at global summits for years, with world leaders always looking for innovative ways to reduce carbon emissions. One promising solution comes from an unlikely source lignin; a natural biopolymer found in plant cell walls. Lignin is a major byproduct of industries like paper production and biofuels. Despite being widely available, it is often overlooked. However, thanks to its high carbon content and resistance to degradation, scientists are starting to recognize lignin as a valuable material for capturing and storing carbon dioxide ($CO_2$). Recent research shows that lignin can be transformed into carbon-rich materials, functionalized polymers, and even biochar, making it a strong candidate for use in carbon capture and storage (CCS) technologies.

### 8.3 Turning lignin into porous carbon for $CO_2$ capture

To use lignin effectively for carbon capture, it must first be converted into porous carbon. These materials are prized for their high surface area, which makes them excellent at trapping gases, such as $CO_2$. A study from the University of Lincoln showed that lignin can be processed into a 3D interconnected carbon structure with large pores ideal for capturing $CO_2$ efficiently [81]. These advanced carbon materials have performed well in tests, showing adsorption capacities that rival conventional materials, such as zeolites and metal-organic frameworks (MOFs).

### 8.4 Improving performance with amine functionalization

Researchers have found that they can improve lignin's $CO_2$ absorbing ability by adding amine group molecules that form strong bonds with carbon dioxide. This process, called amine functionalization, has been a game changer. In one study, industrial lignin was treated with different polyamines, resulting in materials that captured up to 0.80 mmol of $CO_2$ per gram [82]. Another experiment created amine-impregnated mesoporous lignin carbons, which pushed that capture rate even higher—exceeding 3 mmol/g under elevated temperatures [81]. These results show just how much potential lignin has when it is properly engineered.

## 8.5 Lignin-based hydrogels and supraparticles

The innovation does not stop with solids. Scientists have also developed hydrogels and supraparticles made from lignin that work well in more flexible and dynamic settings. One such breakthrough involved embedding amines into lignin-based hydrogels. These hydrogels could capture $CO_2$ quickly and efficiently, and even more impressively, they maintained their performance over multiple cycles of use. In comparison to traditional amine solutions, these materials absorbed more $CO_2$ and could be regenerated with ease—making them a strong candidate for practical, large-scale applications [80].

## 8.6 Direct air capture with ionic lignin polymers

One of the biggest challenges in carbon capture is pulling $CO_2$ directly from the air, where its concentration is very low. This is where ionic lignin polymers come in. Researchers have modified these polymers with quaternary ammonium groups to improve their ability to absorb $CO_2$ from ambient air. These materials have shown the ability to capture over 1 mmol/g from concentrated $CO_2$ sources and about 0.6 mmol/g from regular air all at room temperature. What makes this approach especially promising is that, the captured $CO_2$ can be released on demand and even reused in the production of useful and high value chemicals [83]. It is a powerful combination of sustainability and practicality.

## 8.7 Biodegradation of lignin by microorganisms

### *8.7.1 Key microbial decomposers*

Lignin constitutes a major component of the plant cell wall, accounting for 15–30% of lignocellulosic biomass of cell walls. Its complex and heterogeneous structure, comprising phenylpropanoid units linked by stable ether and carbon-carbon bonds, makes it highly resistant to degradation. Therefore, microbial degradation, especially by fungi and bacteria, presents a sustainable and environmentally friendly alternative to physicochemical methods for lignin depolymerization and valorization [84, 85]. In recent times, researchers have emphasized the metabolic and enzymatic diversity among microbial taxa capable of degrading lignin and their potential applications [86, 87].

### *8.7.2 Fungi-based microbial decomposers*

When considering biodegradation, we can turn to fungi that break down tough plant materials; white-rot fungi, especially types such as *Phanerochaete chrysosporium* and *Trametes versicolor*, are among the most effective at degrading lignin. These fungi release special enzymes, such as lignin peroxidase (LiP), manganese peroxidase (MnP), versatile peroxidase (VP), and laccase [84, 88]. These enzymes help break lignin down into smaller, simpler chemical compounds, such as vanillin and syringaldehyde. Inside the fungal cells, other enzymes then process these compounds further, turning them into useful energy and carbon for growth.

This special ability possessed by the fungi not only helps the fungi decompose wood and plant material but also allows them to break down harmful pollutants in the environment. For example, white-rot fungi can degrade toxic substances, such as dyes from textile factories, polycyclic aromatic hydrocarbons (PAHs) from oil spills, and even some pesticides and industrial chemicals, such as pentachlorophenol (PCP) and polychlorinated biphenyls (PCBs). These qualities make them valuable tools for cleaning up polluted soil and water.

#### *8.7.3 Bacterial-based microbial decomposers*

Compared to fungi, bacteria not as well-studied; nevertheless, certain bacteria are becoming more recognized for their ability to break down lignin. Certain groups of bacteria such as *Actinobacteria, Proteobacteria*, and *Firmicutes* including bacteria such as *Streptomyces*, *Bacillus*, and *Pseudomonas* can degrade lignin using special enzymes like dye-decolorizing peroxidases (DyPs) and bacterial laccases [85, 89].

These bacteria break lignin into smaller aromatic compounds and process them through a series of chemical reactions (called the β-ketoadipate pathway), eventually turning them into energy via the TCA cycle, which is part of their regular metabolism.

Unlike fungal enzymes, bacterial enzymes often work well in a wider range of environmental conditions. For example, they can function at higher temperatures (up to 60°C), in different pH levels (from slightly acidic to alkaline), and in both low-oxygen and oxygen-rich environments [90].

### **8.8 Future prospects and sustainable strategies for lignin**

#### *8.8.1 Advances in biotechnology*

Advances in biotechnology have carved out a future for the use of lignin in this field. Scientists are now modifying microbes such as *Saccharomyces cerevisiae* (a common type of yeast) by adding genes that allow them to break down lignin more effectively. This means these microbes can help turn lignin into valuable products, such as biofuels and chemicals more efficiently [91]. In addition, researchers are also improving enzymes like laccases, proteins that help break down lignin, so they work better under tough industrial conditions. This is done through techniques such as directed evolution, which helps enzymes become more stable and faster at their jobs [92]. These developments can make lignin-based processes more reliable and cost-effective for industries.

#### *8.8.2 Circular economy integration*

Lignin in recent times is also finding exciting new uses in the circular economy, an approach that focuses on reducing waste and reusing materials in a bit to lower cost and improve the environment. For example, lignin is now being added to 3D printing resins to make eco-friendly plastics, and it is being used in green concrete and asphalt to improve durability in construction, while cutting down on fossil-based materials. In Europe, pilot projects are testing lignin-based carbon fibers in the automotive industry to make lightweight car parts, which help improve fuel efficiency. Lignin is also being tested in sustainable construction materials, such as panels and bricks, showing that it can replace less sustainable resources in the building sector.

### *8.8.3 Economic and policy hurdles*

Despite the potential that lignin has, there are still challenges. One of the issues is that fossil-based chemicals are still cheaper and better known in many industries, making it hard for lignin-based products to compete. Another problem one can identify is the lack of clear regulations for lignin-derived materials. For example, Europe has strict certification rules for bio-based products, whereas the U.S., on the other hand, has a more flexible system. These differences in policies and regulations make it difficult for companies to develop and sell lignin-based materials in global markets. Therefore, until policies are more aligned globally and supportive, lignin may struggle to reach its full market potential.

### *8.8.4 Policy recommendations*

For lignin to reach its full potential, governments and industries need to work together. Policy changes should include clearer standards for bio-based materials and financial support for companies developing lignin technologies. Incentives such as tax breaks, subsidies, and government-funded pilot programs can help lower the financial risks for companies investing in lignin. International cooperation to harmonize bio-based product regulations would also create a more predictable market and encourage global investment.

## 9. Conclusion

Lignin, being one of the most abundant polymers on earth, exhibits a remarkably complex and heterogeneous chemical structure, arising from its biosynthetic origin and diverse monomeric composition. This chapter has explored the intricate chemistry of lignin, including its monolignols, polymerization mechanisms, structural difference across plant species, and various inter-unit linkages, which defines its macromolecular architecture.

## Author details

Chiosa Cletus Odili[1*], Anuoluwapo Morenike Ajibola[1], Jesubori Winner Sojobi[1], Omotayo Omolara Ogunbiyi[1], Abiodun Olayinka Ogunsanya[1], Augustine Ben Okoubulu[2], Godwin Omoghegbe-Aigbojie Igbinigie[1], Deborah Osi-Goyeyi Daramola[1], Oludayo Emmanuel Kayode[1], Toluwanimi Samuel Oyegunle[1], Timothy Arome Ayegba[3], Olufemi Tajudeen Salami[1], Olasunkanmi Oriola Akinyemi[1] and Sunday Olayinka Oyedepo[1]

1 Bells University of Technology Ota, Nigeria

2 Southern Delta State University, Ozoro Delta State, Nigeria

3 Centre for Space Transport and Propulsion, National Space Research and Development Agency Federal Ministry of Innovation Science and Technology, Abuja, Nigeria

*Address all correspondence to: ccodili@bellsuniversity.edu.ng

## References


[1] Figueiredo P, Lintinen K, Hirvonen JT, Kostiainen MA, Santos HA. Properties and chemical modifications of lignin: Towards lignin-based nanomaterials for biomedical applications. Progress in Materials Science. 2018;**93**:233-269

[2] Boerjan W, Ralph J, Baucher M. Lignin biosynthesis. Annual Review of Plant Biology. 2003;**54**(1):519-546

[3] Kai D, Tan MJ, Chee PL, Chua YK, Yap YL, Loh XJ. Towards lignin-based functional materials in a sustainable world. Green Chemistry. 2016;**18**:1175

[4] Belgacem MN, Gandini A. Monomers, Polymers and Composites from Renewable Resources. Netherland: Elsevier; 2008

[5] Zhang M, Soyeon J. Green co-solvent-assisted one -pot synthesis of high-performance flexible lignin polyurethene foam. Chemical Engineering Journal. 2024;**499**:156142

[6] Li C et al. Chemical composition and structure of lignin-based polyurethanes. Industrial Crops and Products. 2015;**65**:324-333

[7] Baker DA, Rials TG. Recent advances in low-cost carbon fiber manufacture from lignin. Journal of Applied Polymer Science. 2013;**130**(2):713-728

[8] Qian Y, Qiu X, Zhu S. Lignin: A nature-inspired sun blocker for broad-spectrum sunscreens. Green Chemistry. 2015;**17**(1):320-324

[9] Nair SS et al. High performance green barriers based on lignin and nanocellulose. ACS Sustainable Chemistry & Engineering. 2014;**2**(4):772-780

[10] Houston RW, Abdoulmoumine NH. Investigation of the thermal deconstruction of β–β′ and 4-O-5 linkages in lignin model oligomers by density functional theory (DFT). RSC Advances. 2023;**13**:6181-6190. DOI: 10.1039/D2RA07787F

[11] Bonawitz ND, Chapple C. The genetics of lignin biosynthesis: Connecting genotype to phenotype. Annual Review of Genetics. 2010;**44**:337-363

[12] Ralph J, Lundquist K, Brunow G, Lu F, Kim H, Schatz PF, et al. Lignins: Natural polymers from oxidative coupling of 4-hydroxyphenylpropanoids. Phytochemistry Reviews. 2004;**3**:29-60

[13] Umezawa T. Diversity in lignan biosynthesis. Phytochemistry Reviews. 2003;**2**:371-390

[14] Vanholme R, Demedts B, Morreel K, Ralph J, Boerjan W. Lignin biosynthesis and structure. Plant Physiology. 2010;**153**(3):895-905

[15] Joffres B, Laurenti D, Charon N, Daudin A, Quignard A, Geantet C. Thermochemical conversion of lignin for fuels and chemicals: A review. Oil & Gas Science and Technology - Revue de l IFP. 2013;**68**:753-763. DOI: 10.2516/ogst/2013132

[16] Guadix-Montero S, Sankar M. Review on catalytic cleavage of C-C inter-unit linkages in lignin model compounds: Towards lignin depolymerisation. 2018;**61**(3):183-198

[17] Hatfield RD, Vermerris W. Lignin formation in plants: The dilemma of linkage specificity. Plant Physiology. 2001;**126**(4):1351-1357

[18] Ralph J, Landucci LL. Lignin and Lignans: Advances in Chemistry. USA: CRC Press; 2010. pp. 137-234

[19] del Río JC, Rencoret J, Prinsen P, et al. Structural characterization of lignin from different olive varieties. Journal of Agricultural and Food Chemistry. 2012;**60**(27):7079-7090

[20] Mellerowicz EJ, Sundberg B. Wood cell walls: Biosynthesis, developmental dynamics and their implications for wood properties. Current Opinion in Plant Biology. 2008;**11**:293-300

[21] Barros J, Serk H, Granlund I, Pesquet E. The cell biology of lignification in higher plants a review. Annals of Botany. 2015;**115**(7):1053-1074. DOI: 10.1093/aob/mcv046

[22] Hatfield R, Marita JM. Grass cell walls: A story of cross-linking. Plant Science. 2010;**178**(3):147-156

[23] Balk M, Sofia P, Neffe AT, Tirelli N. Lignin, the lignification process, and advanced, lignin-based materials. International Journal of Molecular Sciences. 2023;**24**(14):11668

[24] Sewalt VJH, Ni W, Blount JW, Jung HG, Masoud SA, Howles PA, et al. Reduced lignin content and altered lignin composition in transgenic tobacco down-regulated in expression of L-phenylalanine ammonia-lyase or cinnamate 4-hydroxylase. Plant Physiology. 1997;**115**:41-50

[25] Kim H, Ralph J, Lu FC, Ralph SA, Boudet AM, MacKay JJ, et al. NMR analysis of lignins in CAD-deficient plants. Part 1. Incorporation of hydroxycinnamaldehydes and hydroxybenzaldehydes into lignins. Organic & Biomolecular Chemistry. 2003;**1**:268-281

[26] Lapierre C, Pollet B, MacKay JJ, Sederoff RR. Lignin structure in a mutant pine deficient in cinnamyl alcohol dehydrogenase. Journal of Agricultural and Food Chemistry. 2000;**48**:2326-2331

[27] Xie M, Zhang J, Tschaplinski TJ, Tuskan GA, Chen J-G, Muchero W. Regulation of lignin biosynthesis and its role in growth-defense tradeoffs. Frontiers in Plant Science. 2018;**9**:1427. DOI: 10.3389/fpls.2018.01427

[28] delRío JC, Rencoret J, Gutiérrez A, Elder T, Kim H, Ralph J. Lignin monomers from beyond the canonical monolignol biosynthetic pathway: Another brick in the wall. ACS Sustainable Chemistry & Engineering. 2020;**8**:4997-5012

[29] Eudes A, Dutta T, Deng K, Jacquet N, Sinha A, Benites VT, et al. SbCOMT(Bmr12) is involved in the biosynthesis of tricin-lignin in sorghum. PLoS One. 2017;**12**:e0178160

[30] Onkokesung N, Gaquerel E, Kotkar H, Kaur H, Baldwin IT, Galis I. MYB8 controls inducible phenolamide levels by activating three novel hydroxycinnamoyl-coenzyme A: Polyamine transferases in Nicotiana attenuata. Plant Physiology. 2012;**158**:389-407

[31] Weng JK, Chapple C. The origin and evolution of lignin biosynthesis. NewPhytologist. 2010;**187**:273-285

[32] Boija E, Johansson G. Interactions between model membranes and lig Nin-related compounds studied by immobilized liposome chromatography. Biochimica et Biophysica Acta. 2006;**1758**:620-626

[33] Barros J, Serrani-Yarce JC, Chen F, Baxter D, Venables BJ, Dixon RA. Role of bifunctional ammonia-lyase in

grass cell wall biosynthesis. Nature Plants. 2016;**2**:16050. DOI: 10.1038/nplants.2016.50

[34] Boija E, Lundquist A, Edwards K, Johansson G. Evaluation of bilayer disks as plant cell membrane models in partition studies. Analytical Biochemistry. 2007;**364**:145-152

[35] Miao YC, Liu CJ. ATP-binding cassette-like transporters are involved in the transport of lignin precursors across plasma and vacuolar membranes. Proceedings of the National Academy of Sciences, USA. 2010;**107**:22728-22733

[36] Tsuyama T, Kawai R, Shitan N, et al. Proton-dependent coniferin trans port, a common major transport event in differentiating xylem tissue of woodyplants. Plant Physiology. 2013;**162**:918-926

[37] Lora JH, Glasser WG. Recent industrial applications of lignin: A sustainable alternative to non-renewable materials. Journal of Polymers and the Environment. 2002;**10**(1-2):39-48

[38] Zhao H, Jones CL, Baker GA, Xia S, Olubajo O, Person VN. Regenerating cellulose from ionic liquids for an accelerated enzymatic hydrolysis. Journal of Biotechnology. 2009;**139**(1):47-54

[39] Higuchi T. Microbial degradation of lignin: Role of lignin peroxidase, manganese peroxidase, and laccase. Proceedings of the Japan Academy, Series B. 2004;**80**(5):204-214

[40] Brebu M, Vasile C. Thermal degradation of lignin—A review. Cellulose Chemistry and Technology. 2010;**44**(9):353-372

[41] Vuković JP, Tišma M. The role of NMR spectroscopy in lignocellulosic biomass characterisation: A mini review. Food Chemistry: Molecular Sciences. 2024;**9**:100219

[42] Pawade SS, Toom L, Herodes K, Leito I. Accuracy of quantitative NMR analysis: A case study of lignin. Journal of Chemical Metrology. 2023;**17**(1):108-121. DOI: 10.25135/jcm.88.2304.2753

[43] Blindheim FH, Ruwoldt J. The effect of sample preparation techniques on lignin Fourier transform infrared spectroscopy. Polymers. 2023;**15**(13):2901. DOI: 10.3390/polym15132901

[44] Ruwoldt J, Tanase-Opedal M, Syverud K. Ultraviolet spectrophotometry of lignin revisited: exploring solvents with low harmfulness, lignin purity, Hansen solubility parameter, and determination of phenolic hydroxyl groups. ACS Omega. 2022;**7**(50):46371-46383

[45] LaVallie A, Andrianova AA, Schumaker J, Reagen S, Lu S, Smoliakova IP, et al. Unfolding of lignin structure using size-exclusion fractionation. Polymers. 2023;**15**(19):3956

[46] Xiang A, Ebdon JR, Horrocks AR, Kandola BK. On the utility of thermogravimetric analysis for exploring the kinetics of thermal degradation of lignins. Bioresource Technology Reports. 2022;**20**:101214

[47] Pe J III, Mun J, Mun S. Thermal characterization of Kraft lignin prepared from mixed hardwoods. BioResources. 2023;**18**(1):926

[48] Belouadah Z, Nasri K, Toubal L. The effects of lignin on the thermal and morphological properties and damage mechanisms after UV irradiation of polypropylene biocomposites reinforced with flax and pine fibres:

Acoustic emission analysis. Materials. 2024;**17**(11):2474

[49] Reyes-Rivera J, Terrazas T. Lignin analysis by HPLC and FTIR: Spectra deconvolution and S/G ratio determination. In: Agusti J, editor. Xylem. Methods in Molecular Biology. Vol. 2722. New York, NY: Humana; 2024. pp. 149-169. DOI: 10.1007/978-1-0716-3477-6_12

[50] Stark NM, Yelle DJ, Agarwal UP. In: Omar F, Mohini S, editors. Techniques for characterizing lignin: Lignin in polymer composites. USA: Elsevier; 2016. pp. 49-66

[51] Ragauskas AJ, Beckham GT, Biddy MJ, et al. Lignin valorization: Improving lignin processing in the biorefinery. Science. 2014;**344**(6185):1246843

[52] Georgs V, Piili H, Gustafsson J, Xu C. A critical review on lignin structure, chemistry, and modification towards utilisation in additive manufacturing of lignin-based composites. Industrial Crops and Products. 2025;**233**:121416

[53] Zakzeski J, Bruijnincx PCA, Jongerius AL, Weckhuysen BM. The catalytic valorization of lignin for the production of renewable chemicals. Chemical Reviews. 2010;**110**(6):3552-3599

[54] Lopez-Camas K, Arshad M, Ullah A. Chemical modification of lignin by polymerization and depolymerization. In: Sharma S, Kumar A, editors. Lignin. Springer Series on Polymer and Composite Materials. Turkey; 2020. pp. 139-180. DOI: 10.1007/978-3-030-40663-9_5

[55] Zhu L, Cui C, Liu H, Zhou Z, Qi F. Thermochemical depolymerization of lignin: Process analysis with state-of-the-art soft ionization mass spectrometry. Frontiers in Chemical Engineering. 2022;**4**:982126. DOI: 10.3389/fceng.2022.982126

[56] Schutyser W, Renders T, Van den Bosch S, et al. Chemicals from lignin: An interplay of lignocellulose fractionation, depolymerisation, and upgrading. Chemical Society Reviews. 2018;**47**(3):852-908

[57] Parsell TH et al. A synergistic biorefinery based on catalytic conversion of lignin prior to cellulose starting materials. Green Chemistry. 2015;**17**(3):1492-1499

[58] Laurichesse S, Avérous L. Chemical modification of lignins: Towards biobased polymers. Progress in Polymer Science. 2014;**39**(7):1266-1290

[59] Kadla JF, Kubo S. Lignin-based polymer blends. Current Opinion in Solid State and Materials Science. 2003;**7**(4-5):281-287

[60] Zhang Y et al. Alkylation of lignin to improve compatibility with polyolefins. ACS Sustainable Chemistry & Engineering. 2016;**4**(2):684-692

[61] Gellerstedt G. Softwood Kraft lignin: Raw material for the future. Industrial Crops and Products. 2015;**77**:845-854. ISSN 0926-6690

[62] Reverchon E. Micro and nano-particles produced by supercritical fluid assisted techniques: Present status and perspectives. Chemical Engineering Transactions. 2002;**2**:1-10

[63] Cocero MJ, Cabeza A, Abad N, Adamovic T, Vaquerizo L, Martinez CM, et al. Understanding biomass fractionation in subcritical & supercritical water. Journal of Supercritical Fluids. 2018;**133**:550-565

[64] Zhang S, Sun J, Zhang X, Xin J, Miao Q, Wang J. Ionic liquid-based green processes for energy production. Chemical Society Reviews. 2014;**43**:7838-7869

[65] Underkofler KA, Teixeira RE, Pietsch SA, Knapp KG, Raines RT. Separation of lignin from corn Stover hydrolysate with quantitative recovery of ionic liquid. ACS Sustainable Chemistry & Engineering. 2015;**3**:606-613

[66] Roy R, Rahman MS, Amit TA, Jadhav B. Recent advances in lignin depolymerization techniques: A comparative overview of traditional and greener approaches. Biomass. 2022;**2**(3):130-154

[67] Smook GA. Handbook for Pulp and Paper Technologists. 4th ed. TAPPI Press; 2016

[68] Vishnuvarthanan M. Sulfite pulping: Extraction. In: Jawaid M, Ahmad A, Meraj A, editors. Handbook of Lignin. Singapore: Springer; 2025. DOI: 10.1007/978-981-97-2664-6_12-1

[69] Fache M, Boutevin B, Vanillin SC. Production from lignin and its use as a renewable chemical. ACS Sustainable Chemistry & Engineering. 2015;**4**:1-13

[70] Effendi H, Gerhauser AV. Bridgwater, production of renewable phenolic resins by thermochemical conversion of biomass: A review. Renewable and Sustainable Energy Reviews. 2008;**12**(8):2092-2116. ISSN 1364-0321. DOI: 10.1016/j.rser.2007.04.008

[71] Jianglei Q, Michael W, Jinwen Z. Use of polycarboxylic acid derived from partially depolymerized lignin as a curing agent for epoxy application. ACS Sustainable Chemistry and Engineering;**2**(2):188-193

[72] Hu F, Ragauskas A. Pretreatment and lignocellulosic chemistry. Bioenergy Research. 2012;**5**(4):1043-1066. DOI: 10.1007/s12155-012-9208-0

[73] Chand S. Review carbon fibers for composites. Journal of Materials Science. 2000;**35**:1303-1313. DOI: 10.1023/A:1004780301489

[74] Okay O. Macroporous copolymer networks. Progress in Polymer Science. 2000;**25**(6):711-779

[75] Kumar R, Mago G, Balan V, Wyman CE. Physical and chemical characterizations of corn Stover and poplar solids resulting from leading pretreatment technologies. Bioresource Technology. 2009;**100**(17):3948-3962. ISSN 0960-8524. DOI: 10.1016/j.biortech.2009.01.075

[76] Ruel-Gariépy E, Leroux J-C. In situ-forming hydrogels—review of temperature-sensitive systems. European Journal of Pharmaceutics and Biopharmaceutics. 2004;**58**(2):409-426. ISSN 0939-6411. DOI: 10.1016/j.ejpb.2004.03.019

[77] Kögel-Knabner I. The macromolecular organic composition of plant and microbial residues as inputs to soil organic matter. Soil Biology and Biochemistry. 2017;**105**:1-8. DOI: 10.1016/j.soilbio.2016.11.005

[78] QingHua F, FanGeng C, HuiRan W. Preparation and characterization of a temperature-sensitive lignin-based hydrogel. BioResources. 2011;**6**(4):4942-4952

[79] Thevenot M, Dignac MF, Rumpel C. Fate of lignins in soils: A review. Soil Biology and Biochemistry. 2010;**42**(8):1200-1211. DOI: 10.1016/j.soilbio.2010.03.017

[80] Moreno A, García J, López F. Monolithic, hybrid and particulate lignin-based hydrogels for sustainable $CO_2$ capture. Green Chemistry. 2024;**26**(3):789-802. DOI: 10.1039/D4GC05489J

[81] Sani M, Ahmad MA, Hassan M. Lignin-based porous carbon adsorbents for $CO_2$ capture. Chemical Society Reviews. 2023;**52**(5):1234-1256. DOI: 10.1039/D4CS00923A

[82] Wang J, Li X, Zhang Y. Amine functionalized lignin-based mesoporous cellular carbons for efficient $CO_2$ capture. Fuel. 2024;**345**:127899. DOI: 10.1016/j.fuel.2023.127899

[83] Ten E, Vermerris W. Recent developments in polymers derived from industrial lignin. Journal of Applied Polymer. 2015;**132**(24):1-13

[84] Torres-Farradá G, Pérez J, Martínez AT. White rot fungi as tools for the bioremediation of xenobiotics: A review. Environmental Microbiology Reports. 2024;**16**(2):123-135. DOI: 10.1111/1758-2229.13045

[85] Wang Y, Chen H, Liu Z. Bacterial transformation of lignin: Key enzymes and high-value products. Biotechnology for Biofuels. 2022;**15**(1):123. DOI: 10.1186/s13068-022-02447-4

[86] Gu Y, Wang X, Li Y. Microbial degradation of lignin: Recent advances and future perspectives. Bioresource Technology. 2024;**356**:127312. DOI: 10.1016/j.biortech.2022.127312

[87] Grgas D, Novak S, Vujčić Z. Lignin biodegradation by bacteria: A review. Journal of Environmental Management. 2023;**302**:113978. DOI: 10.1016/j.jenvman.2021.113978

[88] Kato S, Watanabe T, Nakamura Y. Metabolic mechanism of lignin-derived aromatics in white-rot fungi. Applied Microbiology and Biotechnology. 2024;**108**(2):345-356. DOI: 10.1007/s00253-024-13371-4

[89] Rahayu AP, Putri DA, Nugroho RA. Newly isolated ligninolytic bacteria and its applications for multiple dye degradation. Environmental Science and Pollution Research. 2023;**30**(15):45678-45689. DOI: 10.1007/s11356-023-06377-7

[90] Sumranwanich T, Charoenpanich J, Thongaram T. Evaluating lignin degradation under limited oxygen conditions by bacterial isolates from forest soil. Scientific Reports. 2024;**14**(1):13350. DOI: 10.1038/s41598-024-64237-8

[91] Linger JG, Vardon DR, Guarnieri MT, Karp EM, Hunsinger GB, Franden MA, et al. Lignin valorization through integrated biological funneling and chemical catalysis. Proceedings of the National Academy of Sciences. 2014;**111**(33):12013-12018. DOI: 10.1073/pnas.1410657111

[92] Brissos V, Martins LO, Cosme AM. Engineering laccases for improved stability and activity. Applied Microbiology and Biotechnology. 2017;**101**(10):3711-3721. DOI: 10.1007/s00253-017-8231-5

Chapter 3

# Modelling Lignin: A New Approach to an Old Problem

*Christopher W.J. Murnaghan*


## Abstract

The functionalities which are found within lignin present a rich potential for the cultivation of feedstocks for further processing and synthesis. The linkages and bonding patterns found in lignin are as a result of a totally random process dominated by radical intermediates in the biosynthesis process. Numerous methods have been sought to attempt to degrade and depolymerize lignin but largely these are fraught with difficulties. One approach which has been taken is to synthesise these linkages to form more manageable fragments which can then be employed for catalytic consumption to enable the refinement of operating conditions for the catalyst. This chapter serves as an overview of the key bonding patterns which have been recreated synthetically for the purposes of the development of catalytic systems for the long term goal of valorisation of lignin towards value-added products.


**Keywords:** lignin, organic chemistry, biomass conversion, lignocellulose reforming, catalysis

## 1. Introduction

Lignin finds itself as the most abundant aromatic polymer found in nature, it is composed of three monolignol species which bond together in a range of different linkages of either C−C or C−O provenance [1]. The biosynthetic pathway of lignin relies on the formation of reactive intermediates in the form of radicals which can be delocalized throughout the structure of the monolignol. This means that there are various points on the structure of each monolignol where bonding with another monolignol can occur. This is important when it comes to the bonding patterns (**Figure 1**) which can arise from the result of these radical coupling reactions. With each of the bonding patterns there is a significant difference in the bond enthalpies [2] which must be ultimately considered when it comes to the degradation and conversion of lignin towards value-added compounds.

Lignin is a maze of linkages and its lack of regularity of structure makes the controlled depolymerization very challenging and as a result many strategies have been devised in order to attempt to overcome the conversion of this natural polymer [1, 3, 4]. As it stands, 2% of lignin extracted is used for value-added compounds through conversion processes with the remaining 98% being burnt as a low cost fuel [5].

Ultimately when it comes to the depolymerization of lignin, the problems will always be the same; lack of regularity, the bonding patterns and lack of solubility. As

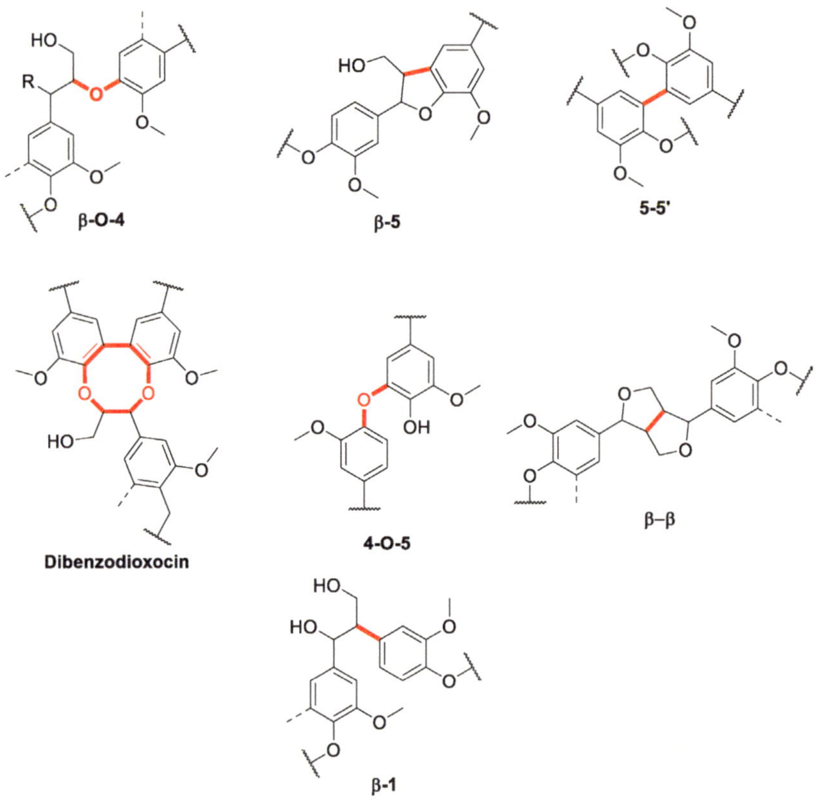


**Figure 1.**
*Range of linkages which are found within softwood and hardwood lignin varieties.*

the bonding patterns have been previously elucidated using advanced NMR [6–8] techniques then building lignin from a 'ground up' approach will give a realistic portion which can be put together to test a range of catalytic systems to determine their effectiveness and potential selectivity.

If one method for depolymerisation is considered; photocatalysis, then the lack of solubility of extracted lignin leads to a highly turbid solution which has issues when light penetration is considered. Even on a wider point, the lack of solubility of lignin extracts gives relatively poor mass transfer due to the lack of homogeneity of the reaction solution. This is a significant issue when the interface at the surface of the catalyst is considered also. As a result, the quest towards finding a system which will give robust breakdown shifts slightly away from the employment of raw lignin and instead an alternative is brought in temporarily for the refinement of such breakdown and depolymerisation strategies.

This gives rise to a new class of research; lignin model chemistry, using synthetic organic chemistry molecules which have similar C—C or C—O bonds like those found within lignin were synthesised and used to test catalytic systems. The rationale behind this bottom up approach is that the catalytic system can be refined and developed fully and then has a better chance of being active against native lignin which has been extracted from raw biomass. Indeed many of the reports which will be described throughout this chapter will have been by researchers who synthesised model compounds solely for the testing of a new depolymerization technology and not simply for synthetic adventure. Throughout this chapter, we will discuss the major and important progresses which have been made in the synthesis of a range of lignin model compounds which contain the linkages displayed in **Figure 1** along with a new class of proposed model compounds. Alongside the successes in synthesis of these model linkages, throughout there will be discussion of limitations of synthesised model compounds and how they are useful to a point.

## 2. β-O-4 containing model

The β-O-4 linkage is the most common in lignin throughout both softwood and hardwood varieties [1], it is an ether type bond which typically contains a hydroxymethyl functionality on the β-carbon. This linkage is by far the most recreated within the literature for a number of reasons; it is relatively simple to make with little synthetic expertise and it will always serve as a good gauge as to whether or not a catalytic system is efficient at cleaving bonds. One of the primary reasons for this is the vulnerability of the β−C−O bond which can be easily cleaved following oxidation of the α-C−O alcohol [2] as can be seen in **Figure 2**. As a result of this phenomenon in the bonding pattern enthalpies, quite a majority of depolymerization and conversion strategies for lignin model compounds has relied on a procedure which will oxidise the −OH at the benzylic position resulting in a weakened β−C−O linkage which will eventually break forming at least two products.

There have been many examples of the synthesis of this type of linkage within the literature both in phenolic and non-phenolic forms to represent the types of lignin which are found in nature [9]. In the typical synthetic route towards a lignin model compound containing this linkage, there is a platform substrate which is usually an acetophenone moiety which may or may not have a *para*-phenol functionality which will require protection. The implementation of this type of linkage as previously mentioned is relatively easily performed and one of the main steps in the process is typically the activation of an aromatic ketone species. In the majority of cases within

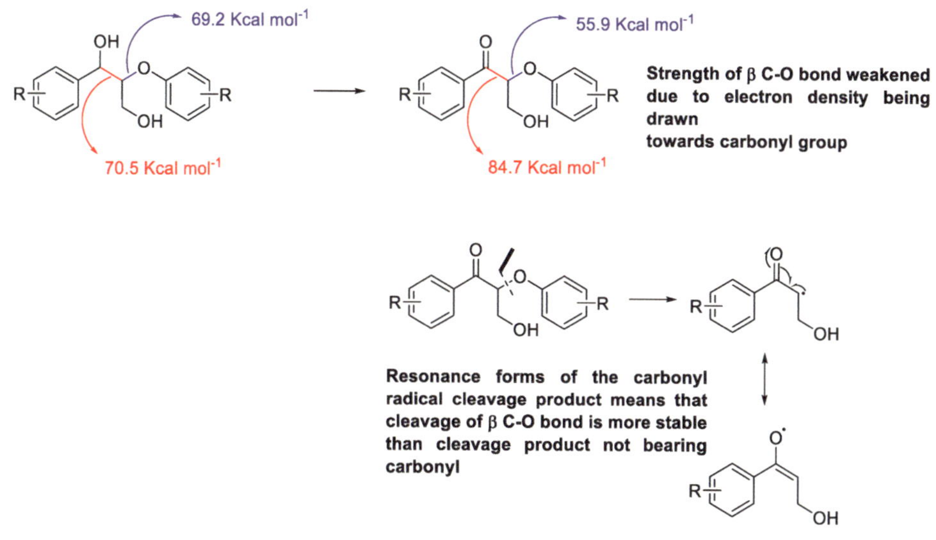


**Figure 2.**
*Oxidation of the benzylic alcohol in the β-O-4 linkage gives rise to a much weaker β−C−O bond which is observed in many degradation studies.*

the literature this is normally done through the use of copper (II) bromide as the brominating agent [10].

Following this, a nucleophile typically in the form of a phenoxide functionality will be reacted with the bromo derivative and the skeleton linkage is formed. Many reports have also included the incorporation of a hydroxymethyl unit at the β-carbon to furnish the complete linkage [11]. An example of how a synthetic scheme for a dimeric model linkage like this is shown below in **Figure 3**. It is of wider importance also to discuss the presence of phenolic functionality when lignin models are being synthesised, the amount of phenolic functionality can range from 10 to 30% [1]. Therefore consideration must be made to whether or not phenolic functionality should be included in the lignin model being synthesised. Synthetically, it is of ease to either include or preclude phenolics through the use of protecting groups strategies (as can be seen in **Figure 3**).

There have been many examples of this synthetic protocol being used previously within the literature and as far back as 1952 [12] where Adler et al. demonstrated the formation of a lignin model dimer, this sparked the beginning of a range of studies where synthetic protocols were developed for the means of recreating this linkage [12–17]. The ease with which this linkage can be put together means that there are plenty of examples of this type of model compound being employed in degradation studies within the literature. As a number of examples, the use of chiral auxiliaries facilitated the synthesis of diastereomerically pure β-O-4 model dimers by Hartwig et al. where the (S,S) diastereomer was initially synthesised and the use of a Mitsunobu inversion gave the second (S,R) diastereomer [18]. In terms of the general form of the synthetic methodology for this linkage, there are two main routes

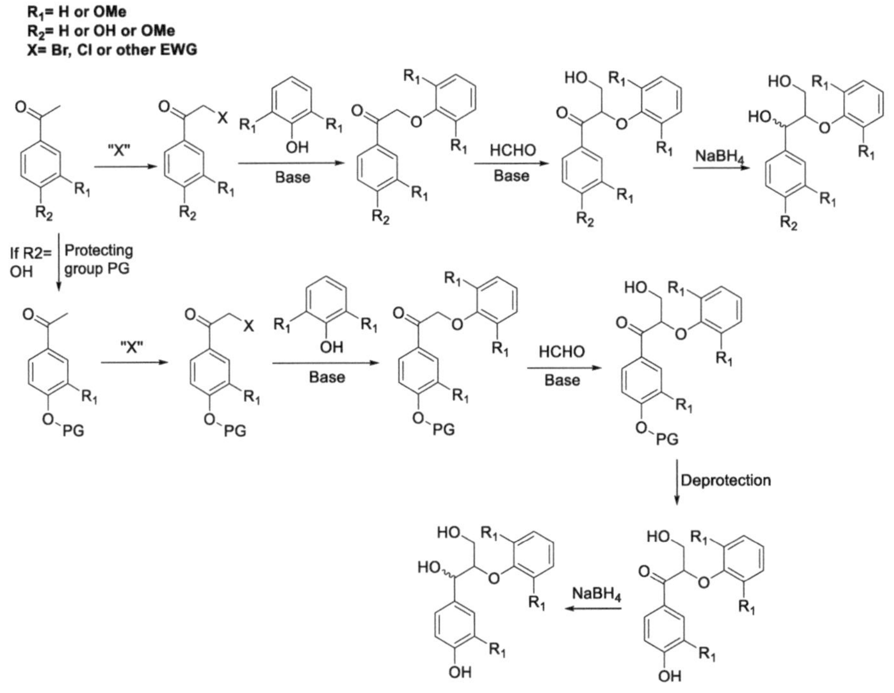


**Figure 3.**
*General synthetic strategy for phenolic and non-phenolic β-O-4 lignin model compounds.*

which have been explored within the literature, the activation of an aromatic ketone typically with a bromide and aldol chemistry. Alongside the implementation of this type of linkage in simple model dimers, this is one of the foremost linkages which is also found in synthetic lignin oligomers and polymers. Through the use of aldol chemistry, formation of a polymeric β-O-4 is much more accessible as the substrate can simply react with itself until the polymer forms in solution. In reality, the synthesis of this type of oligomer [19–22] is much more useful for lignin degradation studies than dimers or trimers as it gives a more representative number of linkages and thus molecular weight which will be closer to a portion of extracted lignin. There is however a wider point to be considered when we discuss β-O-4 model compounds, their respective lack of utility on a larger scale, the fundamental knowledge may be gained at an early stage from a model like this but this in no capacity gives a realistic view of how raw lignin will behave under similar conditions.

Although the biosynthesis of lignin can result in a range of diastereomeric centres being formed, this is a random process and as such the design of a synthetic route which discriminates between enantiomers and diastereomers, although a feat of prowess, can be viewed as being efforts wasted. Such work like that of Hartwig et al. [18, 23] demonstrated that through the implementation of chiral auxiliaries, there can be discrimination in the enantiomeric and diastereomeric products which are received this synthetic route can be seen in **Figure 4**.

This is an important contribution and demonstrates how the use of strategic synthesis can be employed for lignin model compounds to produce compounds for degradation processes to have inherent stereochemistry reflecting somewhat what is observed in nature. Similar efforts were reported by Mukhtar et al. [24] where the inversion of a stereocentre resulted in the formation of a *syn* relationship between the oxygen atoms on the α and β carbon atoms in a β-O-4 skeleton, this can be seen in **Figure 5**.

As a more general point around the synthesis of β-O-4 model compounds, they have been useful to a point and developed the fundamental understanding of many catalytic systems for the conversion of lignin as a wider substrate. The fact however remains that the vulnerable ether linkage found between the β-Carbon and oxygen atom is in fact weakened when the α-carbon is oxidised to provide the benzylic ketone as we have recently found in photocatalytic work [25]. This is a good thing and bad in turn, using this model compound does give good results however can give an inaccurate image of how a catalytic system may react with a larger substrate or one which encompasses more linkages.

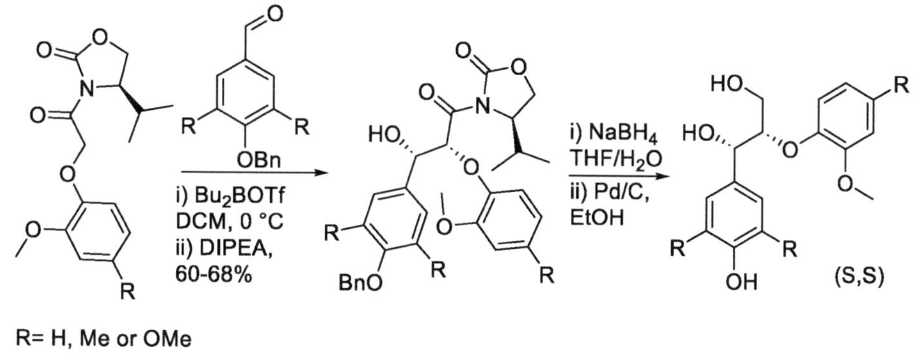


**Figure 4.**
*Implementation of Evans auxiliary for the synthesis of a diastereomeric β-O-4 lignin model compound [18, 23].*

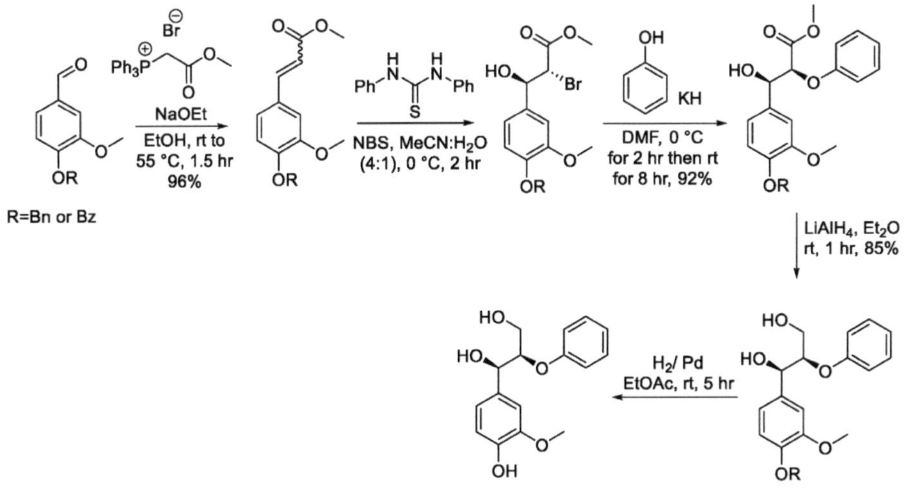


**Figure 5.**
*Mukhtar et al. report of β-O-4 model compound [24].*

## 3. β-5 containing model

The second most common linkage found within lignin is the β-5 linkage which is essentially a benzofuran ring bonded to another aromatic ring, its structure can be seen in **Figure 1**. This model has indeed been synthesised numerous times (**Figure 6**) within the literature however it has been recreated nowhere near as many times as

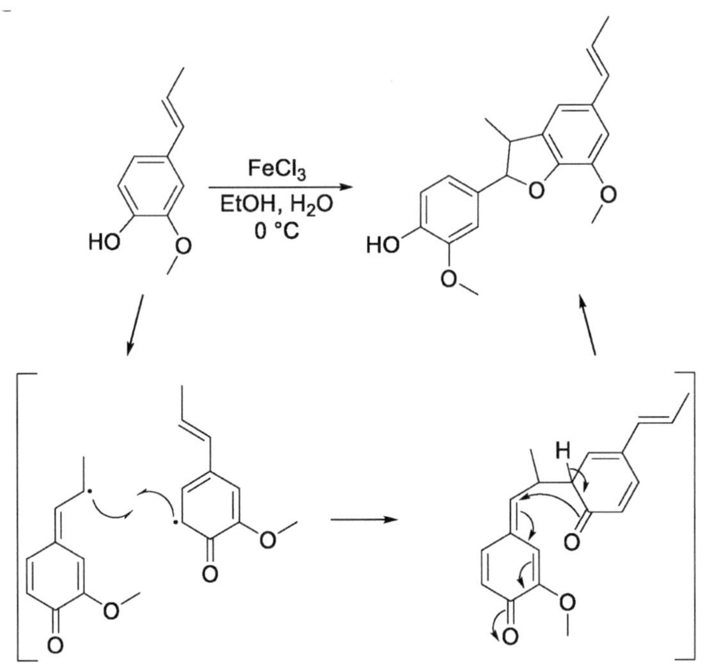


**Figure 6.**
*$FeCl_3$-catalysed oxidative dimerization of isoeugenol to provide dehydrodiisoeugenol.*

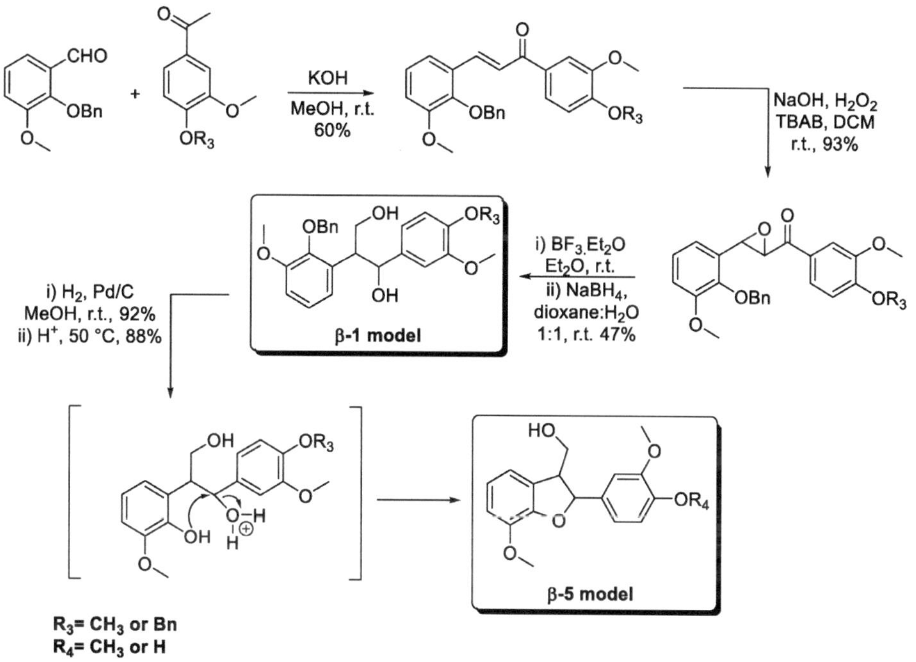


**Figure 7.**
*Synthetic route producing both the β-5 and β-1 linkages.*

that of the β-O-4 linkages. The synthetic methodology which has been employed for the implementation of this linkage ranges from numerous step synthetic routes [26–28] to single step oxidative dimerization processes [29–31]. Some of the original work in the synthetic methodology for the β-5 model compound was done by Leopold Bengt [31] where the oxidative dimerization of isoeugenol furnished dehydrodiisoeugenol this work was later repeated in studies by other workers for various synthetic exploits and lignin model degradation studies [29, 30, 32–34].

This linkage is also an example of which has been produced by enzymatic means within the literature previously, through the use of Horse Radish Peroxidase (HRP) the dimerisation of isoeugenol facilitated the formation of the same structure above dehydrodiisoeugenol. Another strategy was reported also which resulted in the synthesis of two lignin models in a single synthetic route; the β-5 and β-1 [26] demonstrated in **Figure 7**.

With this linkage being the second most common within raw lignin, it is an important one which has been synthesised and explored in terms of degradation studies throughout the literature. There have been a number of other examples of the synthesis of this type of linkage through various enzymatic and traditional chemical transformations [35, 36].

## 4. 5-5′ linkage

The 5-5′ biphenyl linkage is the third most common within lignin and it is also the linkage with the highest bond enthalpy. Similarly to the β-5 linkage, this is not widely recreated in the literature with only a handful of reports demonstrating its formation [37, 38] and

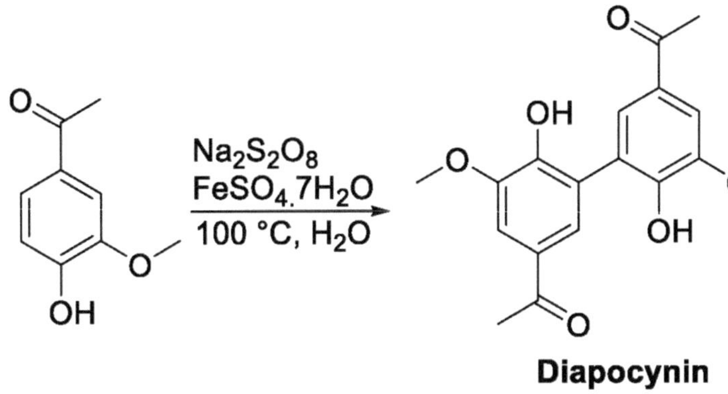


**Figure 8.**
*Oxidative dimerisation of acetovanillone to provide a lignin model compound with a 5-5′ biphenyl linkage.*

subsequent use as a model in lignin chemistry. Looking at this linkage, it seems relatively straightforward to synthesise and is again done through the oxidative dimerization of an aromatic moiety such as acetovanillone or vanillin. Interestingly, one of the model compounds, diapocynin, which has been reported within the literature by Machado et al. also finds use as a pharmaceutically active species which is active against dystrophic muscle [39]. Work in our own lab demonstrated that this model linkage can be easily and in excellent yield synthesised through the use of sodium persulfate and iron sulfate heptahydrate in a one-step protocol (**Figure 8**).

A similar type of model compound was employed by Machado et al. in their investigation into the role of the hydroxyl radical in the cleavage of linkages within lignin [10]. It is worth noting that for conversion studies including this type of linkage, there has been a reduced capacity for many catalytic systems to have the ability to cleave the 5-5′ biphenyl linkage due to the high enthalpy including in our own lab with both enzyme and photocatalysis being unable to cleave the C–C bond [38, 40].

## 5. Other linkages present in lignin

Although these types of linkage are the most prevalent within native lignin, there a range of other model compounds which have been synthesised including other lesser common linkages. These include the β-1, 4-O-5, dibenzodioxocin and β-β and range in their complexity and arrangement. The use of aldol chemistry facilitated the synthesis of the β-1 linkage which was then reduced to afford diastereomeric model dimers with this linkage seen in **Figure 9**.

One of the more complex bonding patterns which is found within lignin is the dibenzodioxocin linkage, which is an eight-membered ring containing two ether bonds. Brunow et al. published numerous reports of the use of both chemical and enzymatic synthesis of model compounds containing this linkage (**Figure 10**) [42, 43]. It appears that the synthesis of this model linkage is one of the more challenging with the chemistry required and the control of reaction. Other linkages including the 4-O-5 and β-β have been synthesised using a range of metal-catalysed and horseradish peroxidase mediated reactions [8, 44]. These linkages are less prevalent in the lignin biopolymer and therefore as such within the literature there are comparatively fewer reports of their syntheses.

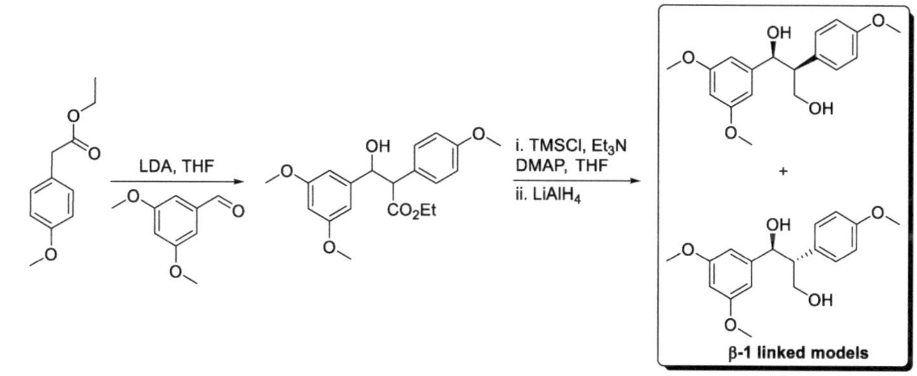


**Figure 9.**
*Synthesis of a β-1 type lignin model compound by Sedai et al. [41].*

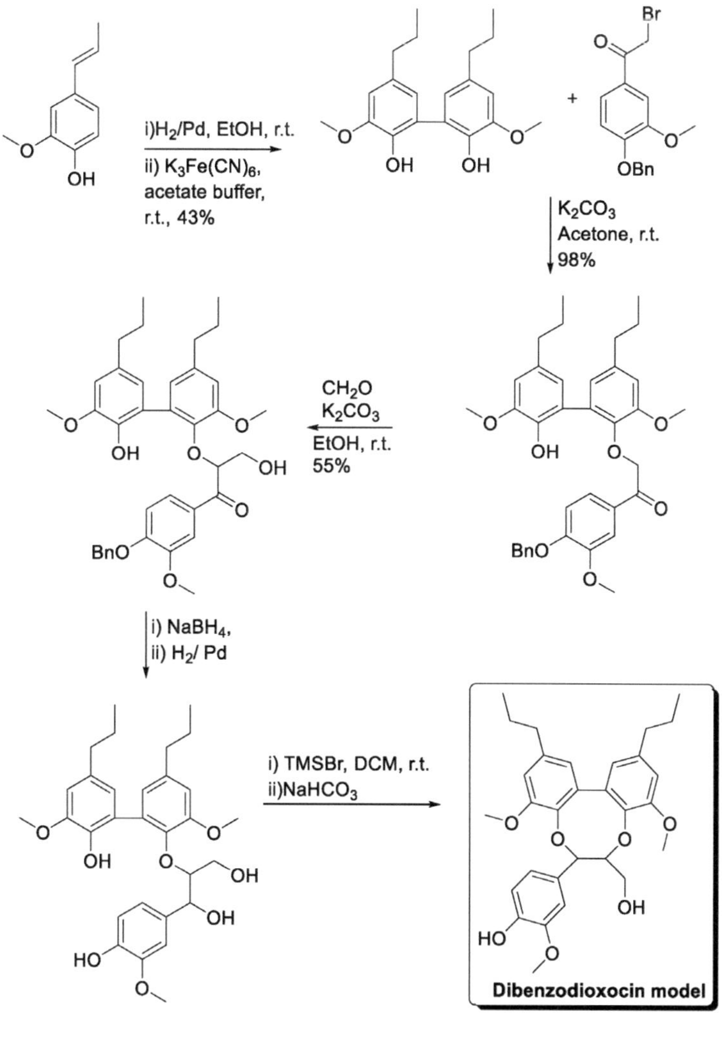


**Figure 10.**
*Synthesis of the dibenzodioxocin model compound by Brunow et al. [42, 43].*

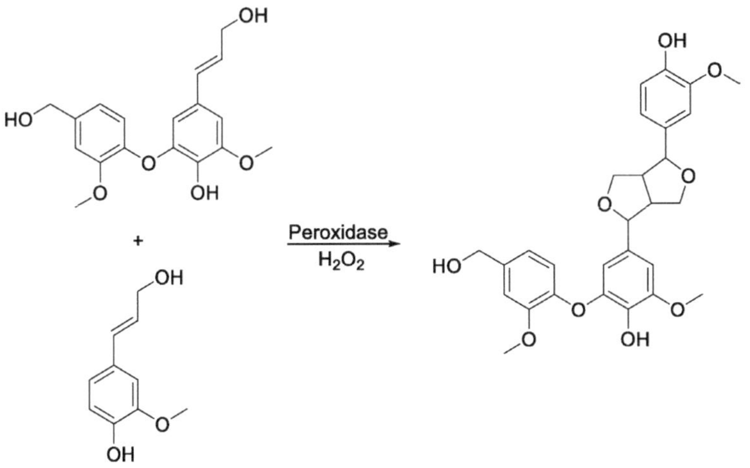


**Figure 11.**
*Synthesis of a trimer by Yue et al. [6, 45] containing the 4-O-5 and β-β linkage.*

The synthetic reports of lignin models of this kind provide an excellent platform for continuing studies into their respective breakdown such as was done in the report by Yue et al. [6, 45]. In this study, the use of horseradish peroxidase was instrumental in the synthesis of model compounds which contained the β-β and also the 4-O-5 linkage (**Figure 11**); both bonding patterns which are not frequently seen within the literature and therefore this work represents an important contribution to the field.

There remains throughout these pieces a central issue with lignin model chemistry thus far in the literature, they have been very well documented in terms of the synthetic prowess and the ease in some circumstances with which the linkages can be synthetically recreated. However the solubility and lack of resemblance on a larger scale remain a central issue with small model dimers, trimers etc. The development of larger polymeric and hybrid molecules will be the stepping stone to which we can find a pathway for the efficient and selective conversion of lignin and biomass towards compounds of interest and added value.

## 6. Models with more than one linkage and different linkages within a single model

As the shift towards producing more reliable catalytic systems for the conversion of lignin is growing, there have been numerous reports of the use of synthetic chemistry for the formation of model compounds which have a number of different linkages. These include the synthesis of polymeric and oligomeric lignin model compounds some with β-O-4 linkages [19, 21]. Westwood et al. produced a polymeric model compound with a β-O-4 linkage through the use of a polymerisation strategy based on aldol addition and subsequent reduction of the ester functionality [20] similar efforts were reported by Katahira et al. [19]. Some pioneering work by Forsythe et al. demonstrated the use of simple chemistry for the synthesis of hexameric and octameric lignin model compounds both containing β-O-4, β-5 and 5-5′ linkages. The development of this synthetic strategy was a major leap forward in the lignin model community and subsequent studies have highlighted the recalcitrance of the linkages found within the hexamer [38, 40, 46, 47]. These model compounds each in their

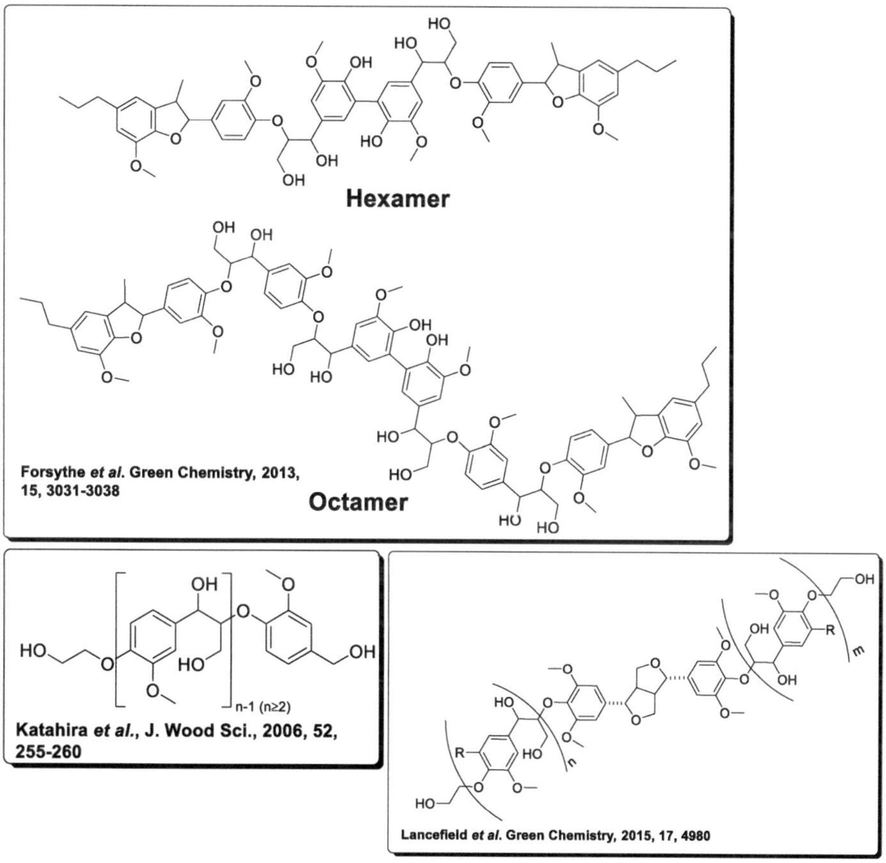


**Figure 12.**
*Range of oligomeric lignin model compounds which have been synthesised within the literature.*

own right represent significant leaps in the research of lignin models for the refinement of catalytic systems. The sheer size and diversity of linkages represents a more diverse substrate for many studies and gives an indication as to how raw lignin might behave under similar conditions. However, the flip side of this is that as the number of aromatic rings increases owing to producing a more representative model compound this largely decreases the aqueous solubility which is an ultimate goal when it comes to the conversion of biomass towards value-added compounds. Below (**Figure 12**) shows some of the different oligomeric model compounds which have been synthesised within the literature.

## 7. Hybrid models

As it is the case within raw biomass, there is a large amount of crosslinking between the three functionalities cellulose, hemicellulose and lignin. Varying techniques have been employed to aid in the fractionation process of lignin from the cellulose and hemicellulose present in biomass [48–51]. This remains a significant advancement in the process of biomass reforming, however it does add a step in the process and in some cases can alter the structure of the extracted fractions [52].

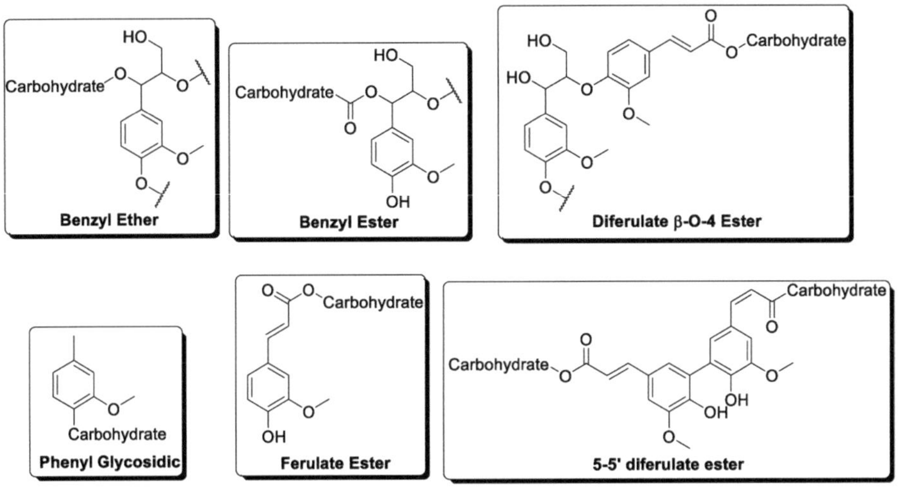


**Figure 13.**
*Different linkages present between lignin and carbohydrate portions of biomass [53].*

As a result of the cross-linking and bonding patterns found between the three entities which have been identified by advanced NMR techniques [53–58]. It has been reported that between lignin and carbohydrate portions take on six main structural motifs each quite individual in their bonding pattern which are shown below in **Figure 13**.

Due to these structures being elucidated directly from biomass sources, they give the most realistic picture of what lignin is actually found like in nature; therefore these are the types of model compounds which must be recreated in order to find refine and develop the best catalytic systems. As it stands within the literature, there are no reports of the conversion of structures like this towards value-added products and such like. This is a niche in the research community and is an area which will see major advancements in the next few years. In terms of the synthetic chemistry which is employed for the formation of some of these linkages, there have been advancements in our own lab and numerous other previous reports. These have centered around the phenyl glycosidic, benzyl ether and ester and have taken numerous approaches to the bonds being formed between the lignin and carbohydrate portion of the molecules. Beginning with our own work [18] the use of both glucose and cellobiose were employed as the 'model' portion of cellulose and xylose was employed for hemicellulose, from there a range of simple lignin models were bonded through the anomeric position of each of the carbohydrate portions, the general scheme of these model compound syntheses can be seen below in **Figure 14**.

The activation of the anomeric position through the use of a trichloroacetimidate species enabled the formation of an O-glycoside species which completed the formation of the protected LCC model compound. The subsequent work which was done and still unpublished using these model compounds has observed that they are fully water soluble which opens up an avenue for new catalytic methodology where previously there had been issues when it came to wholly aromatic substrates in previous breakdown studies. Moving towards some of the other work in this field by Toikka et al. seen the formation of a quinone methide species which is thought to be the reactive intermediate responsible for the biosynthesis of lignin in the formation

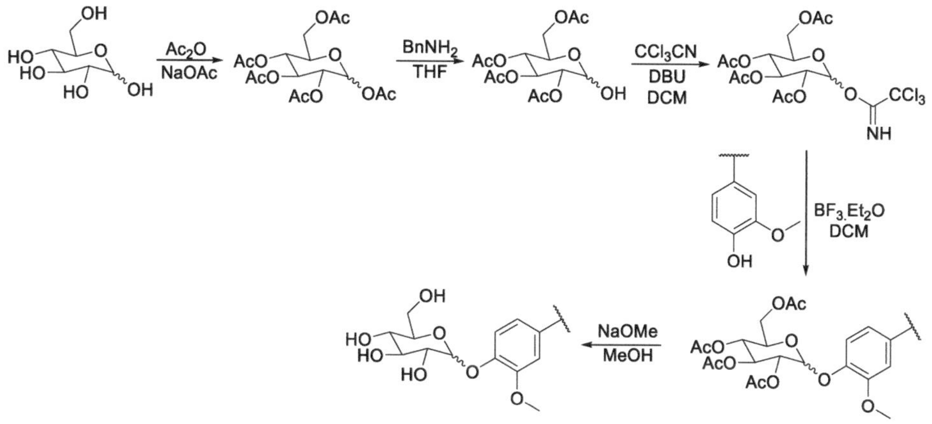


**Figure 14.**
*Route for synthesis of phenyl glycosidic class of lignin carbohydrate complexes by Sheldrake et al.*

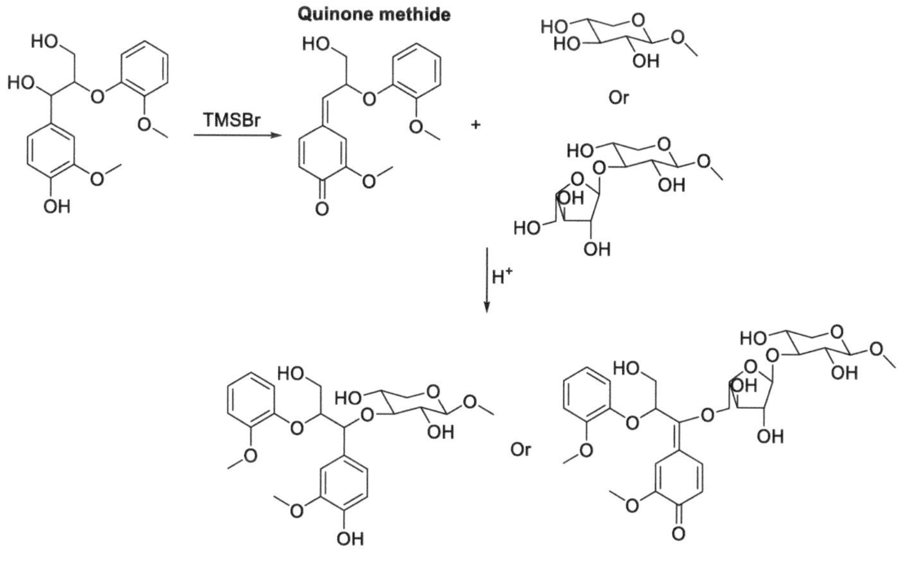


**Figure 15.**
*Synthesis of benzyl ether LCC model compounds by Toikka et al.*

of a majority of linkages. Employing this reactive quinone methide, the formation of LCCs with the benzyl ether linkage were accessible through a simple 1,6-addition reaction (**Figure 15**).

## 8. Future discoveries and advancements

The employment of synthetic organic chemistry for lignin research has indeed been a real gamechanger when it comes to depolymerization of the biopolymer, however it is not without its drawbacks. In no doubt has the implementation of model compounds for the refinement of catalytic systems been important for the lignin chemistry field. However the use of simple models like the β-O-4 linkage is simply not good enough for the focus of catalytic studies, it is now the time to pivot towards model compounds which contain a range of linkages or functionalities which have been explored in the recent sections within this chapter. For example, the use of simple models for a conversion process is theoretically fine until the scale up is considered and there must be thought put towards the plethora of linkages which are present within lignin and the lack of uniformity in 'real' lignin. Another factor which is often poorly considered is solubility, as a general rule and previously discussed, extracted lignin is largely insoluble in a range of solvents which bottlenecks advances in conversion processes. Therefore, as a community, lignin chemists must put serious thought towards moving away from model compounds which have been extensively used throughout the literature for the reporting of a catalytic process. And instead focus must be given to the use of more representative examples of models of biomass which either have more than one type of linkage or a wider array of functionalities. The syntheses of these new type of 'biomass model compounds' can be synthetically more challenging than the previous models however the results and products obtained will more accurately reflect what is obtained when raw biomass is tackled. This will give the community a better chance of being able to tackle the real problem which is the catalytic conversion of lignin towards value-added compounds. It is also an extremely valid view that breaking down lignin is a puzzle that may never be fully cracked due to its recalcitrance and also the major solubility issues. Therefore, this brings to the forefront the importance of representative model compounds which can enable the community to develop strategies which have the ability to selectively target linkages. This can enable us to reach a stage where there can be a specific strategy employed for the yielding of a specific functionality or product which is desired.

## Author details

Christopher W.J. Murnaghan
Department of Biology, Edge Hill University, Ormskirk, UK

*Address all correspondence to: murnaghc@edgehill.ac.uk

## References

[1] Zakzeski J, Bruijnincx PCA, Jongerius AL, Weckhuysen BM. The catalytic valorization of ligning for the production of renewable chemicals. Chemical Reviews. 2010;**110**:3552-3599

[2] Kim S, Chmely SC, Nimlos MR, Bomble YJ, Foust TD, Paton RS, et al. Computational study of bond dissociation enthalpies for a large range of native and modified lignins. Journal of Physical Chemistry Letters. 2011;**2**:2846-2852

[3] Pandey MP, Kim CS. Lignin depolymerization and conversion: A review of thermochemical methods. Chemical Engineering and Technology. 2011;**34**:29-41

[4] Xiang Z, Han W, Deng J, Zhu W, Zhang Y, Wang H. Photocatalytic conversion of lignin into chemicals and fuels. ChemSusChem. 2020;**13**:4199-4213

[5] Christopher LP. Integrated Forest Biorefineries. RSC Publishing; 2012. pp. 1-66

[6] Yue F, Lu F, Ralph S, Ralph J. Identification of 4-O-5-units in softwood lignins via definitive lignin models and NMR. Biomacromolecules. 2016;**17**:1909-1920

[7] Sette M, Wechselberger R, Crestini C. Elucidation of lignin structure by quantitative 2D NMR. Chemistry - A European Journal. 2011;**17**:9529-9535

[8] Tran F, Lancefield CS, Kamer PCJ, Lebl T, Westwood NJ. Selective modification of the β–β linkage in DDQ-treated Kraft lignin analysed by 2D NMR spectroscopy. Green Chemistry. 2015;**17**:244-249

[9] Martínková L, Grulich M, Pátek M, Křístková B, Winkler M. Bio-based valorization of lignin-derived phenolic compounds: A review. Biomolecules. 2023;**13**:717-742

[10] Crestini C, D'Auria M. Singlet oxygen in the photodegradation of lignin models. Tetrahedron. 1997;**53**:7877-7888

[11] Crestini C, D'Auria M. Photodegradation of lignin : The role of singlet oxygen. Journal of Photochemistry and Photobiology A: Chemistry. 1996;**101**:69-73

[12] Adler E. The beta-guaiacyl ether of alpha-veratrylglycerol as a lignin model. Svensk Pepperstidn. 1952;**55**:245-254

[13] Adler E, Eriksoo E. Guaiacylglycerol and its beta-guaiacyl ether. Acta Chemica Scandinavica. 1955;**9**:341-342

[14] Lundquist SRK. New synthetic routes to lignin model compounds of the arylglycerol-beta-aryl ether type. Acta Chemica Scandinavica. 1975;**B29**:276-278

[15] Hosoya KKHKJNS. Synthesis of Guaiacylglycerol-beta-guaiacyl ether. Mokuzai Gakkaishi. 1980;**26**:118-121

[16] Katayama FNTHT. Syntheses of arylglycerol-beta-aryl ethers. Mozukai Gakkaishi. 1981;**27**:223-230

[17] Miksche GE. Zur Synthese der beiden diasteromeren Formen des Guajacylgly. Acta Chemica Scandinavica. 1966;**20**:1038-1943

[18] Njiojob CN, Bozell JJ, Long BK, Elder T, Key RE, Hartwig WT. Enantioselective syntheses of lignin models: An efficient synthesis of β-O-4 dimers and trimers by using the Evans

chiral auxiliary. Chemistry - A European Journal. 2016;**22**:12506-12517

[19] Katahira R, Kamitakahara H, Takano T, Nakatsubo F. Synthesis of β-O-4 type oligomeric lignin model compound by the nucleophilic addition of carbanion to the aldehyde group. Journal of Wood Science. 2006;**52**:255-260

[20] Lancefield CS, Westwood NJ. The synthesis and analysis of advanced lignin model polymers. Green Chemistry. 2015;**17**:4980-4990

[21] Ciofi-Baffoni S, Banci L, Brandi A. Synthesis of oligomeric mimics of lignin. Journal of the Chemical Society-Perkin Transactions 1. 1998;**2**:3207-3217

[22] Chu S, Subrahmanyam AV, Huber GW. The pyrolysis chemistry of a β-O-4 type oligomeric lignin model compound. Green Chemistry. 2013;**15**:125-136

[23] Njiojob CN, Rhinehart JL, Bozell JJ, Long BK. Synthesis of enantiomerically pure lignin dimer models for catalytic selectivity studies. Journal of Organic Chemistry. 2015;**80**:1771-1780

[24] Mukhtar A, Zaheer M, Saeed M, Voelter W. Synthesis of lignin model compound containing a β-O-4 linkage. Zeitschrift fur Naturforschung - Section B Journal of Chemical Sciences. 2017;**72**:119-124

[25] Sheldrake GN, Skillen N, Robertson PKJ, Murnaghan CWJ. Photocatalytic conversion of β-O-4 lignin model dimers: The effect of benzylic ketones on reaction pathway. Catalysts. 2025;**15**:525-567

[26] Fang Z, Meier MS. Toward the oxidative deconstruction of lignin: Oxidation of β-1 and β-5 linkages. Organic & Biomolecular Chemistry. 2018;**16**:2330-2341

[27] Brunow G, Lundquist K, Carlström K, Sköldefors H, Wilking N, Theve NO. A new synthesis of model compounds for the beta-5 structural unit in lignins. Acta Chemica Scandinavica. 1984;**38b**:335-336

[28] Li S, Lundquist K. Synthesis of lignin models of β-5 type. Acta Chemica Scandinavica. 1997;**51**:1224-1228

[29] Forsythe WG, Garrett MD, Hardacre C, Nieuwenhuyzen M, Sheldrake GN. An efficient and flexible synthesis of model lignin oligomers. Green Chemistry. 2013;**15**:3031

[30] Mardhatillah A, Mutakin M, Levita J. FeCl 3 -catalyzed synthesis of dehydrodiisoeugenol. International Journal of Chemistry. 2016;**8**:1-7

[31] Leopold B, Hassel O, Lund EW. Aromatic keto- and hydroxy-polyethers as lignin models. III. Acta Chemica Scandinavica. 1950;**4**:1523-1537

[32] Sheldrake GN, Curran NM, Murnaghan CWJ. Synthetic efforts towards the phenyl glycosidic class of lignin-carbohydrate complex models. Biomass and Bioenergy. 2025;**193**:107555. DOI: 10.1016/j.biombioe.2024.107555

[33] Liu J, Ralphs K, Murnaghan C, Skillen N, Sheldrake G, McCarron P, et al. Exploring the photocatalytic cleavage pathway of the β-5 linkage lignin model compound on carbon nitride. ChemSusChem. 2025;**18**(3):e202400955. DOI: 10.1002/cssc.202400955

[34] Murnaghan CWJ, Skillen N, Hardacre C, Bruce J, Sheldrake GN, Robertson PKJ. Exploring lignin valorisation : The application of photocatalysis for the degradation of

the β-5 linkage. Journal of Physics: Energy. 2021;**3**:1-10

[35] Lahive CW, Lancefield CS, Codina A, Kamer PCJ, Westwood NJ. Revealing the fate of the phenylcoumaran linkage during lignin oxidation reactions. Organic & Biomolecular Chemistry. 2018;**16**:1976-1982

[36] Ouyang XP, Yang Y, Zhu GD, Qiu XQ. Radical synthesis of tetrameric lignin model compound. Chinese Chemical Letters. 2015;**26**:980-982

[37] Machado AEH, Furuyama AM, Falone SZ, Ruggiero R, da Silva Perez D, Castellan A. Photocatalytic degradation of lignin and lignin models, using titanium dioxide: The role of the hydroxyl radical. Chemosphere. 2000;**40**:115-124

[38] Murnaghan CWJ, Skillen N, Hackett B, Lafferty J, Robertson PKJ, Sheldrake GN. Toward the photocatalytic valorization of lignin: Conversion of a model lignin hexamer with multiple functionalities. ACS Sustainable Chemistry & Engineering. 2022;**10**:12107-12116

[39] Ismail HM, Scapozza L, Ruegg UT, Dorchies OM. Diapocynin, a dimer of the NADPH oxidase inhibitor apocynin, reduces ROS production and prevents force loss in eccentrically contracting dystrophic muscle. PLoS One. 2014;**9**:1-8

[40] Murnaghan CWJ, Forsythe WG, Lafferty JH, Sheldrake GN. Biocatalytic conversion of lignin model oligomer using a laccase-mediator system. Green Chemistry. 2024;**26**:10851-10858

[41] Sedai B, Díaz-Urrutia C, Baker RT, Wu R, Silks LAP, Hanson SK. Aerobic oxidation of β-1 lignin model compounds with copper and oxovanadium catalysts. ACS Catalysis. 2013;**3**:3111-3122

[42] Karhunen P, Rummakko P, Pajunen A, Brunow G. Synthesis and crystal structure determination of model compounds for the dibenzodioxocine structure occurring in wood lignins. Journal of the Chemical Society, Perkin Transactions 1. 1996;**4**:2303-2308

[43] Karhunen P, Rummakko P, Sipilä J, Brunow G, Kilpeläinen I. The formation of dibenzodioxocin structures by oxidative coupling. A model reaction for lignin biosynthesis. Tetrahedron Letters. 1995;**36**:4501-4504

[44] Lahive CW, Deuss PJ, Lancefield CS, Sun Z, Cordes DB, Young CM, et al. Advanced model compounds for understanding acid-catalyzed lignin depolymerization: Identification of renewable aromatics and a lignin-derived solvent. Journal of the American Chemical Society. 2016;**138**:8900-8911

[45] Yue F, Lu F, Sun R, Ralph J. Synthesis and characterization of new 5-linked pinoresinol lignin models. Chemistry - A European Journal. 2012;**18**:16402-16410

[46] Guadix-Montero S, Sainna MA, Jin J, Reynolds J, Forsythe WG, Sheldrake GN, et al. Ruthenium ion catalysed C-C bond activation in lignin model compounds - Towards lignin depolymerisation. Catalysis Science & Technology. 2023;**13**:5912-5923

[47] Kishimoto T, Uraki Y, Ubukata M. Chemical synthesis of β-O-4 type artificial lignin. Organic & Biomolecular Chemistry. 2006;**4**:1343-1347

[48] Zhan Q, Lin Q, Wu Y, Liu Y, Wang X, Ren J. A fractionation strategy of cellulose, hemicellulose, and lignin from wheat straw via the biphasic pretreatment for biomass valorization. Bioresource Technology. 2023;**376**:128887

[49] D'Orsi R, Di Fidio N, Antonetti C, Raspolli Galletti AM, Operamolla A. Isolation of pure lignin and highly digestible cellulose from defatted and steam-exploded Cynara cardunculus. ACS Sustainable Chemistry & Engineering. 2023;**11**:1875-1887

[50] Montgomery JRD, Bazley P, Lebl T, Westwood NJ. Using fractionation and diffusion ordered spectroscopy to study lignin molecular weight. ChemistryOpen. 2019;**8**:601-605

[51] Gigli M, Crestini C. Fractionation of industrial lignins: Opportunities and challenges. Green Chemistry. 2020;**22**:4722-4746. DOI: 10.1039/d0gc01606c

[52] Maltari R, Kontro J, Koivu K, Farooq M, Mikkilä J, Zhang R, et al. Fractionation of technical lignin from enzymatically treated steam-exploded poplar using ethanol and formic acid. ACS Applied Polymer Materials. 2022;**4**:9388-9398

[53] Tarasov D, Leitch M, Fatehi P. Lignin-carbohydrate complexes: Properties, applications, analyses, and methods of extraction: A review. Biotechnology for Biofuels. 2018;**11**:1-28

[54] Björkman A. Lignin and lignin-carbohydrate complexes extraction from wood meal with neutral solvents. Industrial and Engineering Chemistry. 1957;**49**:1395-1398

[55] Xie Y, Yasuda S, Wu H, Liu H. Analysis of the structure of lignin-carbohydrate complexes by the specific 13C tracer method. Journal of Wood Science. 2000;**46**:130-136

[56] Balakshin M, Capanema E, Berlin A. Isolation and Analysis of Lignin-Carbohydrate Complexes Preparations with Traditional and Advanced Methods: A Review. 1st ed. Vol. 42. Elsevier B.V.; 2014

[57] Du X, Li J, Gellerstedt G, Rencoret J, Del Río JC, Martínez AT, et al. Understanding pulp delignification by laccase-mediator systems through isolation and characterization of lignin-carbohydrate complexes. Biomacromolecules. 2013;**14**:3073-3080

[58] Miyagawa Y, Mizukami T, Kamitakahara H, Takano T. Synthesis and fundamental HSQC NMR data of monolignol β-glycosides, dihydromonolignol β-glycosides and p-hydroxybenzaldehyde derivative β-glycosides for the analysis of phenyl glycoside type lignin-carbohydrate complexes (LCCs). Holzforschung. 2014;**68**:747-760

Section 2

# Emerging Applications of Lignin

Chapter 4

# Advances in Lignin-Based Hybrid Nanomaterials as a Sustainable Approach for Water Treatment

*Anny Leudjo Taka*


## Abstract

Lignin-based hybrid nanomaterials, with their multifunctional and sustainable nature, are emerging as promising materials for a wide range of applications, including energy, water treatment, biomedicine, and catalysis. This chapter will comprehensively review the different approaches to functionalizing lignin and its application in water treatment. The chapter will particularly highlight recent advances in synthesizing lignin-based hybrid nanomaterials such as lignin-based nanoparticles, functionalized lignin nanocomposites, and functionalized lignin polymer nanocomposites. These materials, serving as nano-adsorbent filtration materials, are at the forefront of the battle against organic pollutants (e.g., microplastics), inorganic pollutants (e.g., mercury metal ions), and microorganism contaminants in water, which will also be discussed. The challenges, such as structural variability and factors influencing their contaminant removal capacity, regeneration efficiency, and scalability, will be discussed to guide the future development of high-performance lignin-based hybrid nanomaterials for water purification.


**Keywords:** functionalized lignin, nanocomposites, nano-adsorbents, antimicrobial agent, water purification, microplastics

## 1. Introduction

Nowadays, there is an increasing attention on utilizing sustainable resources, particularly lignocellulosic biomass, for both material and energy purposes. This lignocellulosic biomass is mainly comprised of three biopolymers: lignin, hemicellulose, and cellulose. Among these three biopolymers, lignin is one of the most abundant biopolymers (natural polymers) resulting from renewable resources (e.g., lignocellulosic biomass from the wood or paper pulp industry) [1].

Lignin is generally brownish in color and has a highly branched, amorphous, and heterogeneous structure. Lignin is made by crosslinking three main precursor monolignols (phenolic precursors). The three main phenolic precursors that crosslink are paracoumaryl alcohol (also known as 4-hydroxyphenyl), sinapyl alcohol, and coniferyl alcohol (**Figure 1**). Lignin is insoluble in water but soluble in alkaline and

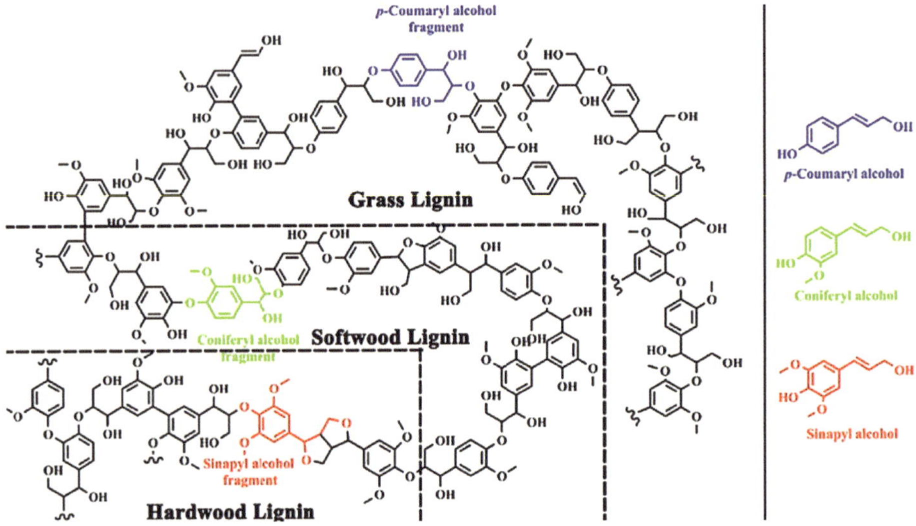


**Figure 1.**
*The origin of lignin, its different structures, and the three main phenolic precursors. Reproduced with permission from Ref. [2].*

specific organic solvents; it exhibits good thermal stability and antioxidant activity. It is more hydrophobic than hemicellulose and cellulose [1, 2]. Lignins, despite their structural complexity, possess various functional groups (e.g., carbonyl, hydroxyl), making them ideal candidates for chemical modification with various types of chemical moieties and nanomaterials. This leads to the creation of advanced lignin-based hybrid nanocomposite materials that exhibit multifunctionality and enhanced performance, sparking their potential in multiple applications, including water treatment [1, 3].

This chapter presents an overview of the functionalization of lignin, highlighting recent advances in synthesizing lignin-based polymer nanocomposites. These nanocomposites, often referred to as "lignin-based hybrid nanomaterials," are the result of modifying lignin's chemical structure to enhance its properties and expand its potential applications. Additionally, the use of these lignin-based hybrid nanocomposites as nano-adsorbents for the removal of organic pollutants (e.g., microplastics), inorganic contaminants (e.g., mercury metal ions), and microbial contaminants in wastewater is also discussed. The challenges, such as extraction, solubility, and structural variability, as well as factors influencing their capacity to remove pollutants from water (e.g., contact time, initial concentration, and pH), regeneration efficiency, and scalability, are also addressed to guide the future development of high-performance lignin-based hybrid nanomaterials for water purification.

## 2. Extraction, solubility, and functionalization of lignin

### 2.1 Extraction and solubility of lignin

Lignin's extraction from lignocellulosic biomass was found to be a complex process because of the intricacy of lignin's structure. Additionally, the properties of lignin obtained after extraction primarily depend on the source of the plant or wood material and the extraction method used [4]. As reported in the literature, various approaches can be used for the extraction of lignin, such as sulfur-free processes (solvent and soda pulping methods) [2], sulfur processes (sulfite and kraft pulping methods) [2, 5], alkaline treatment [6], acid hydrolysis, enzymatic hydrolysis [7], and deep eutectic solvent methods [1, 2, 4]. Among these extraction approaches, the two main processes most widely employed are the sulfur processes and the sulfur-free processes (**Table 1**). In sulfur processes, sulfur-based chemicals are used to facilitate the extraction of lignin using either sulfite pulping or kraft pulping methods. On the other hand, sulfur-free processes, which include the use of either soda or organosolv methods, extract lignin using sulfur-free chemicals. However, the types of lignin obtained from these sulfur-free processes have a low molecular weight and low dispersity, and their structures are more similar to the native lignin [2, 8].

Furthermore, lignin is a complex, branched polymer, and its solubility in many solvents has been demonstrated to be difficult, mainly due to the hydrogen bonding forces, polarity, and dispersion of lignin in common organic solvents (e.g., tetrahydrofuran, DMSO, and methanol). However, among organic solvents, deep eutectic solvents (DES) have been recognized since 2003 as substitute green solvents because of their lower economic and environmental impact [9]. DES can be defined as a mixture containing a hydrogen bond donor (HBD) (e.g., glycerol) and a hydrogen bond acceptor (HBA) (e.g., choline chloride) for which the eutectic point temperature is below that of an ideal liquid mixture [10, 11]. Wang et al. demonstrated the mechanism of DES for dissolving lignin based on Hansen's solubility theory. Their results showed that the best DES solvents

| Extraction processes | Sulfite pulping | Kraft pulping | Soda pulping | Solvent pulping |
|---|---|---|---|---|
| Source of raw material | Hardwood or softwood | Hardwood or softwood | Annual plants | Hardwood, softwood or annual plants |
| Extraction experimental conditions | pH = 1–2, T = 140°C, $HSO_3^-$ | pH = 13, T = 170°C, mixture of $NaOH+Na_2S$ | pH = 11–13, T = 150–170°C, NaOH | Mixture of acetic acid/formic acid/water or only ethanol T = 150–200°C |
| Type of lignin extracted and properties | Lignosulfonate lignin; Soluble in water; sulfur content 4–8%; molecular weight average 15,000–50,000 g/mol; dispersity 6–8; presenting outstanding emulsifying and binding properties. | Kraft lignin; hydrophobic; soluble in alkali organic solvents (because of the deprotonation of phenolic and carboxylic groups); sulfur content 1–3%; molecular weight average 1000–3000 g/mol; dispersity 2.5–3.5 | Soda lignin (or Alkali lignin) Soluble in alkali solvents; molecular weight average 800–3000 g/mol; dispersity 2.5–3.5; high phenolic content because of aryl ether bonds cleavage; solubility decreases under neutral or acidic pH due to reduced ionization of the functional groups. | Organosolv lignin; Soluble in various organic solvents; molecular weight average 500–5000 g/mol; dispersity 1.5–2.5 |

**Table 1.**
*The different approaches to extract lignin from lignocellulosic biomass and the properties of each type of lignin extracted [1, 2].*

for dissolving hydrothermal and alkaline lignin were in the order of G-DES (choline chloride:glycerol), L-DES (choline chloride:lactic acid), M-DES (choline chloride:malic acid), A-DES (choline chloride:glacial acetic acid), and O-DES (choline chloride:oxalic acid) [9]. It is important to note that reaction temperature can also influence lignin solubility in DES, leading to increased solubility at higher temperatures [11]. Therefore, finding the correct green solvent for dissolving lignin plays an important part in its further use in chemical reactions, for example, its modification using a variety of chemical moieties to produce the desired product in higher yield with excellent properties.

## 2.2 Functionalization of lignin

Lignin is a versatile material, and its functionalization means modifying its chemical structure to enhance its properties for various applications. According to the literature, it is also possible to use lignin in its pristine form without further modifications. However, previous studies have demonstrated that modifying lignin is an essential step during synthesis, before its use in advanced processes or reactions [2]. This is because the modified lignin possesses improved properties and additional functional groups valuable for the efficient use of lignin in diverse applications. In general, the three main possible routes for the modification of lignin are (i) the incorporation of new functional, active sites on the structure of lignin; (ii) the use of various chemical reagents to modify the hydroxyl group present on the structure of lignin; and (iii) depolymerization or fragmentation of lignin to yield carbon materials and compounds rich in aromatic groups [2].

In this chapter, we will primarily focus on the approach to modifying lignin by creating new chemically active sites on the lignin structure and chemically modifying the hydroxyl groups present in the lignin structure. This is because the modified lignin obtained *via* these two pathways is highly reactive for their use in many synthetic polymeric composite reactions [2, 3]. In particular, the functionalization of lignin can be accomplished through a rich range of methods, including chemical, physical, or biological [3, 12, 13]. The most widely utilized method is chemical modification, involving the reaction of lignin with chemical moieties of interest through methods such as graft copolymerization, carboxymethylation, phosphorylation, oxidation, sulfonation, esterification, alkylation, amination, and demethylation. Physical approaches to modify lignin include freeze-drying, plasma treatment, irradiation, ultrasonic homogenization, and sorption [2]. Biological methods involve the use of enzymatic degradation and the conversion of lignin into valuable compounds through biological processes [13]. **Table 2** presents a summary of this rich and complex field of lignin functionalization methods.

### *2.2.1 Sulfonation of lignin*

Sulfonation modification, the process of introducing sulfomethyl or sulfonic acid groups onto polymers, holds great promise. It has the potential to considerably increase the stability and dispersion of materials in solutions. The modified lignin obtained *via* sulfonation typically exhibits enhanced chemical stability and hydrophilicity, paving the way for a wide range of applications [3].

### *2.2.2 Demethylation and alkylation of lignin*

Functionalization of lignin by demethylation is a method of subtracting methyl groups from lignin, which occupy the active sites on the aromatic rings, leading to a

| Methods of modifications | Type of reaction processes | Agent used to achieve modification |
|---|---|---|
| Chemical modifications | Methylolation | Aldehydes |
| | Hydrolysis | Chlorosulfonic acid<br>Hydrochloric acid |
| | Alkoxylation | Cyclic ethers |
| | Epoxidation | Epichlorohydrin derivatives |
| | Silylation | Silica-containing agents |
| | Arylation and alkylation | Carboxylic acids<br>Heterocyclic hydrocarbons<br>Chlorinated hydrocarbons |
| | Amination | Amine compounds |
| | Esterification | Carboxylic acid<br>Carboxylic anhydrides<br>Acyl chlorides |
| | Copolymerization | Anilines<br>Lactides<br>Lactones |
| Physical modifications | — | UV irradiation<br>Ultrasonic homogenization and sorption of compounds onto the surface<br>Gamma irradiation<br>Freeze-drying<br>Plasma treatment |

**Table 2.**
*Summary of the different approaches to functionalize lignin [2].*

reduction in the molecular mass of lignin and conversion to other functional groups that improve lignin properties and reactivity. On the other hand, modification of lignin *via* alkylation involves the insertion of alkyl groups into the lignin, increasing the length of the carbon chain. The functionalized lignin obtained *via* this method exhibits improved stability and solubility, making it favorable for various applications [2].

### *2.2.3 Phosphorylation of lignin*

Prieur and et al. [14] have successfully demonstrated the phosphorylation of kraft lignin. This was achieved through the reaction of kraft lignin with phosphorus pentoxide ($P_2O_5$), using tetrahydrofuran as the solvent for the reaction. The reaction was heated to solvent reflux temperature for 8 hours. The phosphorylation mainly involves grafting phosphorus moieties (phosphate groups) to aliphatic hydroxyl groups and phenolic groups of the lignin. Prieur and coworkers have also reported that using $P_2O_5$ to phosphorylate Kraft lignin also resulted in dehydration and decarboxylation of the lignin structure [14].

### *2.2.4 Graft copolymerization of lignin*

In this modification process, lignin reacts with a monomer *via* a polymerization reaction under specific conditions, allowing lignin to gain new properties by choosing suitable monomers, controlling the density and length of the chain, and adjusting the initiation conditions to achieve a lignin-based polymer composite with specific properties [2].

*2.2.5 Carboxymethylation*

Carboxymethylation modification of lignin involves introducing carboxymethyl functional groups into the lignin structure. This modification has shown that carboxymethylated lignin can form complexes with metals through chelation. The chelating property of carboxymethylated lignin makes it a promising candidate for environmental applications, particularly in water treatment for the removal of pollutants [15].

## 3. Advances in the synthesis of lignin-based polymer nanocomposites and their application in water treatment

Access to clean water for sanitation and safe drinking is limited mainly due to pollution and climate change, which is posing a threat to easing the supply of clean water to the growing population. Therefore, there is a need to develop sustainable and multifunctional materials with high surface area and greater affinity for water contaminants, which can be used in water purification processes. In this regard, lignin is known as a low-cost, effective adsorbent, which offers a green and scalable approach for developing multifunctional lignin-based nanocomposite materials that can be applied to remove different classes of water pollutants from contaminated water [1, 16].

Lignin is known as a low-cost, effective adsorbent. However, natural lignin is less reactive and has a limited adsorption site; hence, from the previous section, we have highlighted the importance of functionalizing lignin to improve its properties toward its use as an adsorbent material for water purification. Literature has demonstrated that modified lignin can further be functionalized with other materials through chemical modification to obtain lignin-based nanocomposites or lignin-based polymer nanocomposites, also called lignin-based hydrogels [1, 2].

To date, the development of lignin-based nanocomposites and lignin-based nanoparticles has attracted tremendous attention. These lignin-based nanomaterials have been successfully utilized as adsorbents mainly for the uptake of organic and inorganic (metal ion) pollutants from contaminated water and used as antimicrobial agents [1, 16–21]. The potential future applications of these lignin-based nanomaterials are vast, ranging from large-scale industrial water treatment to point-of-use household water purification systems, and their continued development and optimization could significantly contribute to addressing the global water crisis [1, 16].

The factors influencing the capacity of these lignin-based nanomaterials to remove contaminants from water were mainly reported to be the initial concentration of the pollutants, adsorbent dose, pH, and contact time. The effect of pH was reported to be the key factor affecting the adsorption efficiency of these lignin-based nanomaterials. In fact, at a pH lower than the pH-point-of-zero charge, the adsorption was generally low, whereas the adsorption capacity or percent removal of pollutants was higher at a pH higher than the pH-point-of-zero charge. Additionally, the efficiency of desorption and regeneration of the used adsorbent also depends on the pH of the desorbing solution [16–22]. Regarding the effect of contact time, the adsorption percentage increases as the contact time also increases until the equilibrium is achieved. In addition, an increase in adsorbent dose leading to an increase in sorption capacity was reported to be due to the presence of free adsorption sites at higher adsorbent dosages. Prior studies have also demonstrated that as the initial concentration of pollutant increases, the adsorption capacity also increases until there is a saturation of adsorption active sites by pollutants (adsorbates), where no further increase in adsorption can be observed [16–21].

Furthermore, previous studies have also investigated the mechanism of removal of pollutants by lignin-based nanomaterials using various isotherm and kinetic models, as well as the results obtained based on the study of the effect of solution pH. The mechanism of adsorption was mainly depicted to be a combined process of electrostatic interactions, hydrogen bonding, π-π interactions, ion exchange, hydrophobic interactions, and pore filling (**Figure 2**) [1]. Throughout the mechanism of adsorption, π-electron donors and acceptors from aromatic groups link together *via* π-π interactions. At the same time, the anionic or cationic contaminants are trapped by the oppositely charged functional groups on the adsorbent surface *via* electrostatic interaction. Then, the hydrogen bonding interactions take place between the pollutants and the functional groups (which act as hydrogen acceptors/donors) present on the adsorbent active adsorption site. To reduce contact with water molecules, water contaminants having nonpolar groups tend to agglomerate in water because of hydrophobic interactions with absorbents. On the other hand, *via* the ion exchange mechanism, the anion/cation present on the surface of the adsorbent is replaced by respective anionic/cationic pollutants, leading to metal complexation. Then, for the adsorbents with a high pore volume and surface area, the mechanism of adsorption of pollutants also involves the trapping and diffusion of adsorbate molecules (pollutants) inside the pores of adsorbents *via* pore-filling processes [1, 23].

In this section, the interest mainly centered on the use of lignin-based hybrid nanomaterials as nanosorbent filters for the removal of microplastics, mercury metal ions, and microorganisms from contaminated water. It is also important to note that for the past 10 years, very little research has reported on the synthesis of lignin-based nanocomposites applied as adsorbents to remove microplastics and mercury metal ions, as well as microorganism pollutants, from wastewater.

### 3.1 Removal of organic pollutants: Microplastics

Microplastics (plastic particles less than 5 mm), resulting from the breakdown of waste plastics, are a rising concern because of their detrimental effects on the environment, aquatic biota, and human health. Their tiny size and complex characteristics make them difficult to remove. However, recent studies have shown that lignin-based adsorbent materials offer a sustainable and efficient solution for this problem, providing hope for the future of microplastic removal [24–26].

For example, Gageanu et al. [27] successfully extracted lignin using an ethanol-based organosolv treatment method and a pressurized stirred-tank reactor. The extracted lignin was exploited to adsorb various microplastics such as expanded polystyrene (EPS), polystyrene (PS), polypropylene (PP), and high-density polyethylene (HDPE). This successful work by Gageanu et al. serves as a proof of concept, demonstrating the effectiveness of lignin in removing microplastics [27].

The study by Liu and et al. [15] has demonstrated that functionalizing lignin and further modifying it with other nanomaterials to obtain lignin-based nanocomposites (or lignin-based hydrogels) helps enhance the properties of lignin and increase the

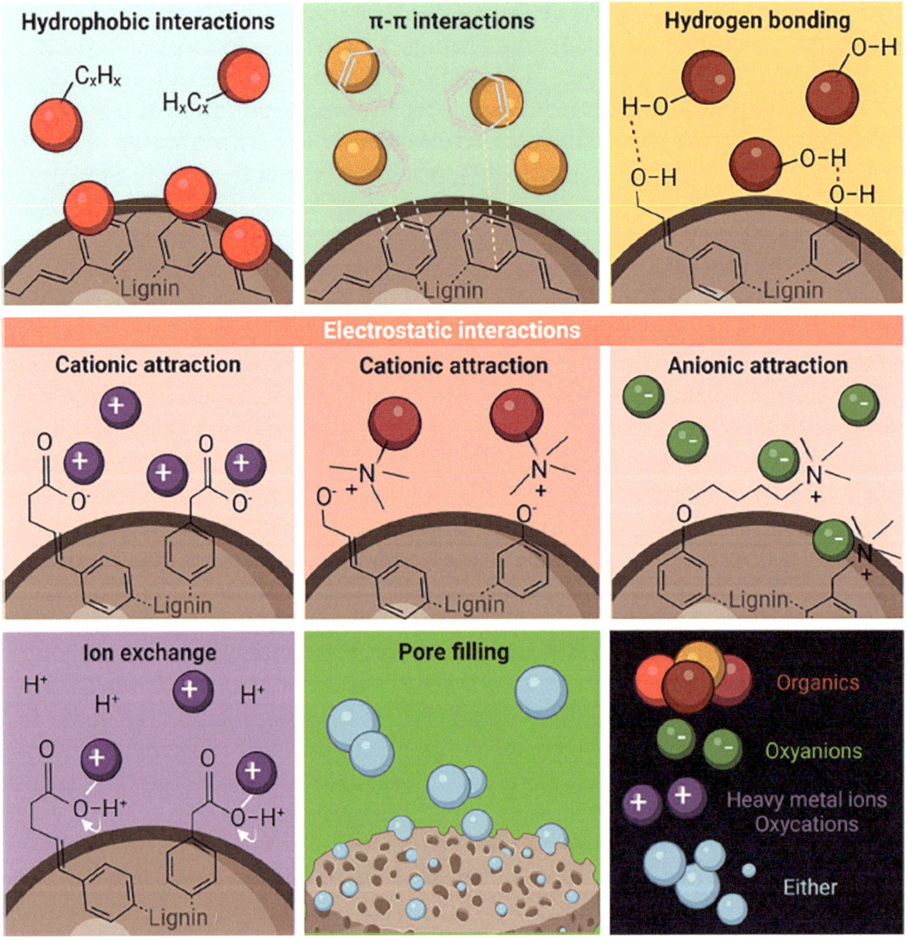


**Figure 2.**
*The different mechanisms used by functionalized lignin-based nanomaterials to adsorb or remove pollutants from water. Reproduced with permission from Ref. [1].*

availability of functional groups at the adsorption site of the lignin-based nanocomposite. Therefore, Liu and coworkers have synthesized a carboxymethylated lignin cuprous oxide nanocomposite (**Figure 3**) to degrade microplastics (polystyrene microspheres) from contaminated water photocatalytically. The CML@$Cu_2O$ showed a performance of 99.68% removal efficiency, compared to 86.88% when only $Cu_2O$ was used. They have also demonstrated that the stability and the reusability of CML@ $Cu_2O$ were maintained during the reaction [15].

Bahi et al. [28] synthesized a lignin-zeolite composite nanofiber membrane using the electrospinning technique. In their work, 1–5 wt% of zeolite nanoparticles were loaded into lignin electrospun nanofibers. According to their reported results, the incorporation of 1 wt% zeolite nanoparticles into lignin electrospun nanofiber membranes has demonstrated enhanced hydrophilicity, mechanical properties, permeability, and separation factor during the purification of contaminated water containing microplastics. Additionally, they have shown that adding more than 1 wt% zeolite nanoparticles to lignin electrospun nanofibers decreases the mechanical properties, hydrophilicity, and permeability [28].

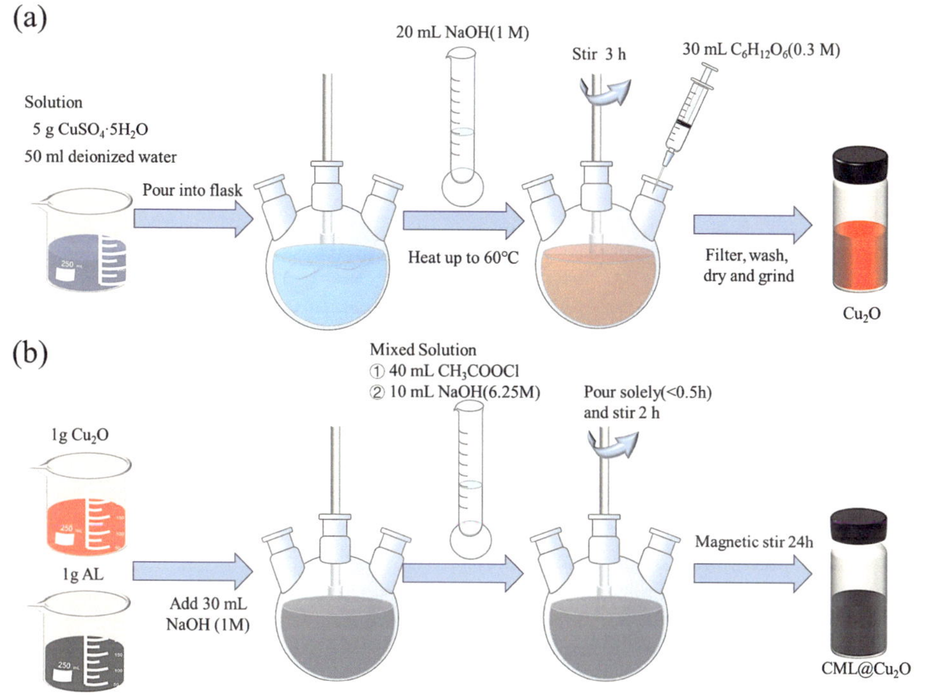


**Figure 3.**
*Synthesis of $Cu_2O$ and carboxymethylated lignin@$Cu_2O$. Reproduced with permission from Ref. [15].*

## 3.2 Removal of inorganic pollutants: Mercury metal ions

Mercury (Hg) is classified as a toxic metal ion according to the Canadian Environmental Protection Act of 1999 [29]. These toxic metal ions are mostly released to wastewater, and they originate from sewers, industrial discharges, wastes from human activities, and agricultural runoff. Mercury can also be found in fish and shellfish. These metal ions are detrimental to aquatic biota, the environment, and human health [29]. For this reason, there is a need to develop eco-friendly, sustainable, recyclable, and efficient materials capable of removing these heavy metal pollutants from contaminated water or wastewater treatment plant effluent to acceptable levels. In this regard, lignin-based nanocomposites have been developed and tested for the removal of Hg from contaminated water [16, 30–32]. However, the research studies on the uptake of Hg from water using lignin-based nanocomposites are limited. Mercury is mainly known as one of the most toxic heavy metal contaminants in aquatic ecosystems because of its non-biodegradability and persistence.

Moreover, Zhou and et al. [19] have developed an amino-modified alkali lignin through the one-step Mannich reaction under mild conditions (**Figure 4a**). The sample was thoroughly characterized using Fourier-transform infrared (FTIR) spectroscopy, scanning electron microscopy (SEM), energy-dispersive X-ray spectroscopy (EDS), and the Brunauer-Emmett-Teller (BET) method for surface area determination. The team demonstrated that the synthesized amino-functionalized alkali lignin (AAL) is highly efficient in removing $Hg^{2+}$ from contaminated water, with an impressive adsorption capacity of 553.9 mg/g. They also showed that the modified AL can be regenerated and reused, maintaining a removal efficiency of over 91% after the 5th cycle of adsorption-desorption. The best desorbing solvent was a mixture of HCl (1 mol/l) and thiourea (1.5%). The reaction mechanism indicates that desorption occurs *via* an ion exchange process in the HCl solution, while acid thiourea enhances ion exchange and forms a stable complex with $Hg^{2+}$ [19]. This efficiency makes the amino-modified alkali lignin a promising candidate for practical applications in mercury removal.

Zhou et al. [33], *via* Friedel-Crafts alkylation reaction, have synthesized a mercapto-functionalized alkali lignin (AL-SH). The synthesized AL-SH was efficient toward the uptake of $Hg^{2+}$ with a maximum adsorption capacity of 101.2 mg/g (**Figure 4b** and **c**). Factors such as the effect of adsorbent dosage, contact time, and pH were investigated. The adsorption was mainly pH dependent, with an optimum pH in the range of pH 4–6. The mechanism of adsorption was best explained by the chemical adsorption process, which was spontaneous and endothermic [33].

In 2020, Dan Zhang et al. [17] prepared a novel polyethyleneimine-functionalized chitosan-lignin composite sponge. This sponge can selectively and efficiently remove Hg (II) ions from aqueous solutions. What is more, it achieves a high removal percentage of over 83.5% within just 1 minute of the optimum adsorption. This quick response time, assigned to the evenly interconnected porous structure of the composite sponge with nanoscale-wall structures, is a significant practical advantage.

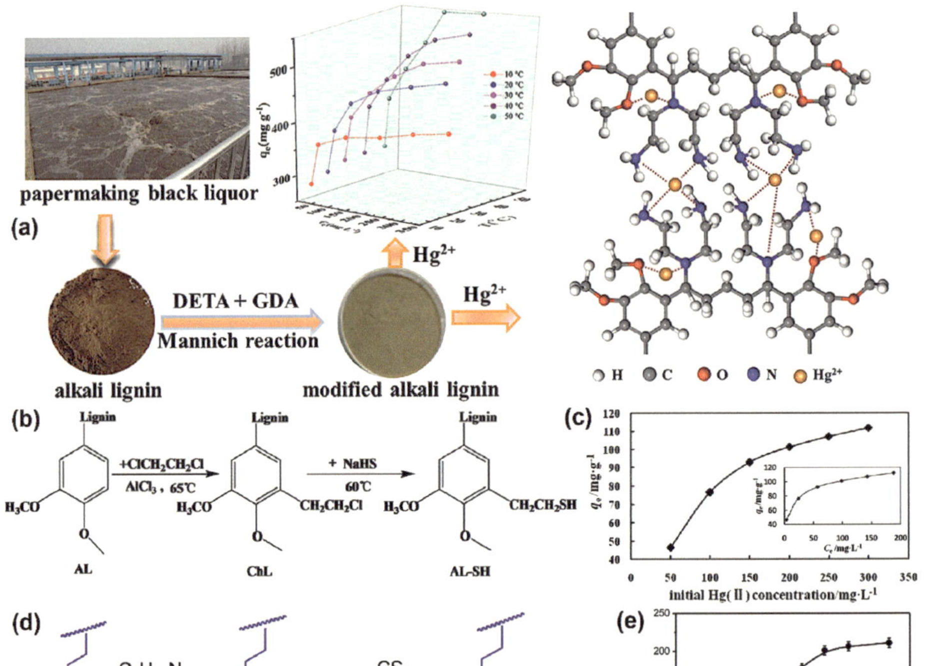


**Figure 4.**
*Illustration of the preparation of mercapto-functionalized alkali lignin (AL-SH) (a and b) and their application in the removal of* $Hg^{2+}$ *(c). Reproduced with permission from Ref. [30].*

The structure increases the distribution of functional groups, providing numerous active sites for solute transportation, thereby increasing the reaction rate and chelation of Hg metal ions by the functional groups present on the surface. The developed polyethyleneimine-functionalized chitosan-lignin composite sponge has been proven to be an effective and practical approach for the sorption of heavy metal ions from the aqueous environment [17].

### 3.3 Removal of microorganism contaminants in wastewater

Wastewater is also a pool of pathogenic microorganism pollutants (e.g., bacteria, fungi, and viruses), posing a high risk to diverse ecosystems and human health. To resolve this problem, conventional disinfection methods such as ultraviolet irradiation, chlorination, and ozonation have been used; however, they have presented some limitations related to the formation of disinfection by-products, high operational costs, or insufficient removal efficiency under varying water quality conditions. As a sustainable and innovative alternative, lignin-based polymer nanocomposites have gained attention for their promising antimicrobial activity, environmental compatibility, and scalability [34].

Literature has shown that lignin possesses phenolic functional groups favorable for its use in the disinfection of contaminated water. Moreover, when lignin is functionalized or hybridized with chemical moieties or nanomaterials, its activity performance is even more enhanced toward a broad spectrum of antimicrobial effects through mechanisms such as the generation of reactive oxygen species (ROS) and cell membrane disruption. In particular, the mechanism of microbial inhibition by lignin-based nanomaterials involves the process of adsorption, whereby the microorganisms initially adhere to the lignin-based nanomaterial surface *via* hydrophobic or electrostatic interactions. This is followed by inactivation processes, which involve the release of metal ions and the generation of ROS, leading to oxidative stress and the damage of cellular components, such as nucleic acids, proteins, and lipids (**Figure 5**) [34, 35].

For instance, Abla Alzagameem et al. [36] demonstrated in their experimental studies the antimicrobial activity of lignin-chitosan composites, lignin-derived cellulose, and lignin against particular spoilage and microorganisms. The antimicrobial activities of functionalized lignins such as hydroxypropylmethylcellulose/lignin (HPMC/lignin) and HPMC/lignin/chitosan demonstrated high antimicrobial activities against Gram-negative and Gram-positive bacteria at 35°C and low temperature (0–7°C). The lignin added to synthesized films (HPMC/lignin and HPMC/lignin/chitosan) produced a positive antimicrobial activity, while the incorporation of chitosan enhanced the antibacterial activity even more. Additionally, a 30% addition of lignin in HPMC/lignin expressed activity against both *P. fluorescens* and *B. thermosphacta*, whereas the addition of only 5% lignin in HPMC/lignin/chitosan showed much better activity against *P. fluorescens* and *B. thermosphacta* [36].

Furthermore, the use of silver nanoparticles (Ag NPs) in various antimicrobial studies has raised serious concerns about environmental toxicity, as active particles of Ag ions are released into the environment after their use, posing a hazard to human health and other ecosystems [37]. To resolve this problem, Richter *et al.* [38] developed lignin-core-Ag NPs, which limit the intrinsic toxicity of Ag while ensuring good attachment of the nanohybrids to bacterial cell membranes, thereby killing the bacteria's broad spectrum. The prepared lignin-core-Ag NPs exhibited biocidal activity using 10 times less Ag than when using conventional Ag NPs. Moreover, Richter and colleagues have demonstrated that the prepared

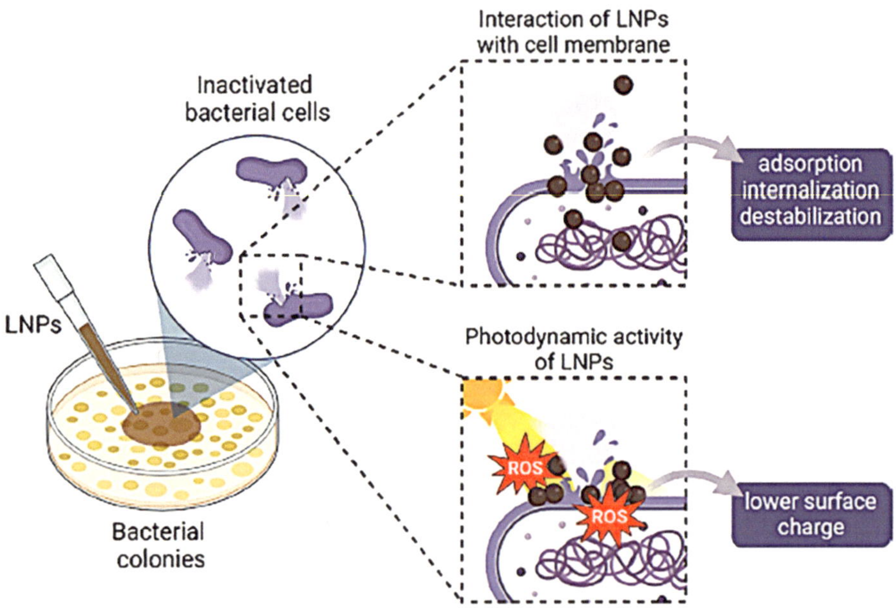


**Figure 5.**
*Picture illustrating the antibacterial activity mechanism of lignin-based nanoparticles (LNPs). Reproduced with permission from Ref. [1].*

lignin-core-Ag NPs lose their antimicrobial activity after use. Hence, it is possible to safely dispose of and degrade used lignin-core-Ag NPs [38].

## 4. Conclusion and further recommendations

The advancement in the development of lignin-based hybrid nanomaterials has been addressed. The different approaches to extract and functionalize lignin have been reviewed and discussed. These lignin-based hybrid nanomaterials have been proven to be versatile materials, possessing multifunctional groups that are valuable for removing various classes of pollutants, including organic, inorganic, and microbial pollutants, from wastewater. Lignin-based hybrid nanomaterials have emerged as promising and sustainable materials that can be used for water treatment. Nonetheless, it is essential to note that there is still limited research on the synthesis of lignin-based nanocomposites or lignin-based polymer nanocomposites (or lignin-based hydrogels) for the removal of microplastics (or nanoplastics) and mercury metal ions, as well as microorganism pollutants from wastewater. Moreover, another challenge that remains is to investigate how to integrate these lignin-based hybrid materials into existing technologies used in water treatment plant effluents while ensuring economic and environmental sustainability.

## Acknowledgements

The authors would like to acknowledge the University of Regina for the financial support and facilities provided.

*DOI: http://dx.doi.org/10.5772/intechopen.1012237*

## Author details

Anny Leudjo Taka
Department of Chemistry and Biochemistry, University of Regina, Regina, Canada

*Address all correspondence to: anny.leudjo@uregina.ca

## References

[1] Camargos CHM, Yang L, Jackson JC, Tanganini IC, Francisco KR, Ceccato-Antonini SR, et al. Lignin and nanolignin: Next-generation sustainable materials for water treatment. ACS Applied Bio Materials. 2025;**8**:2632-2673

[2] Komisarz K, Majka TM, Pielichowski K. Chemical and physical modification of lignin for green polymeric composite materials. Materials. 2023;**16**:1-40

[3] Mengqi S, Shoujuan W, Yuebin X, Weijun Y, Xuliang L, Xueqing Q, et al. Preparation of chemically modified lignin and its application in advanced functional materials. Resources Chemicals and Materials. 2025;**4**:100104

[4] Sheridan E, Filonenko S, Volikov A, Sirviö JA, Antonietti M. A systematic study on the processes of lignin extraction and nanodispersion to control properties and functionality. Green Chemistry. 2024;**26**:2967-2984

[5] Argyropoulos DDS, Crestini C, Dahlstrand C, Furusjö E, Gioia C, Jedvert K, et al. Kraft lignin: A valuable, sustainable resource, opportunities and challenges. ChemSusChem. 2023;**16**:e202300492

[6] Kim JS, Lee YY, Kim TH. A review on alkaline pretreatment technology for bioconversion of lignocellulosic biomass. Bioresource Technology. 2016;**199**:42-48

[7] Watkins D, Nuruddin M, Hosur M, Tcherbi-Narteh A, Jeelani S. Extraction and characterization of lignin from different biomass resources. Journal of Materials Research and Technology. 2015;**4**:26-32

[8] Arapova OV, Chistyakov AV, Tsodikov MV, Moiseev II. Lignin as a renewable resource of hydrocarbon products and energy carriers (a review). Petroleum Chemistry. 2020;**60**:227-243

[9] Wang N, Xu B, Wang X, Lang J, Zhang H. Insights into the mechanism of lignin dissolution via deep eutectic solvents by using Hansen solubility theory. Journal of Molecular Liquids. 2022;**366**:120294

[10] Hladnik L, Vicente FA, Novak U, Grilc M, Likozar B. Solubility assessment of lignin monomeric compounds and organosolv lignin in deep eutectic solvents using in situ Fourier-transform infrared spectroscopy. Industrial Crops and Products. 2021;**164**:113359

[11] Chen H, Wang A, Yan C, Liu S, Li L, Wu Q, et al. Study on the solubility of industrial lignin in choline chloride-based deep eutectic solvents. Sustainability (Switzerland). 2023;**15**:7118

[12] Taleb F, Ammar M, Ben MM, Ben SR, Moussaoui Y. Chemical modification of lignin derived from spent coffee grounds for methylene blue adsorption. Scientific Reports. 2020;**10**:1-13

[13] Zhao L, Zhang J, Zhao D, Jia L, Qin B, Cao X, et al. Biological degradation of lignin: A critical review on progress and perspectives. Industrial Crops and Products. 2022;**188**:115715

[14] Prieur B, Meub M, Wittemann M, Klein R, Bellayer S, Fontaine G, et al. Phosphorylation of lignin: Characterization and investigation of the thermal decomposition. RSC Advances. 2017;7:16866-16877

[15] Liu L, Liu B, Lu Y, Jin X, Chen B, Wang C, et al. Bandgap engineered carboxymethylated lignin cuprous oxide nanocomposites for enhanced visible light photocatalytic degradation of microplastics. Applied Surface Science. 2025;**687**:162278

[16] Ge Y, Li Z. Application of lignin and its derivatives in adsorption of heavy metal ions in water: A review. ACS Sustainable Chemistry & Engineering. 2018;**6**:7181-7192

[17] Zhang D, Wang L, Zeng H, Rhimi B, Wang C. Novel polyethyleneimine functionalized chitosan-lignin composite sponge with nanowall-network structures for fast and efficient removal of Hg(II) ions from aqueous solution. Environmental Science: Nano. 2020;**7**:793-802

[18] Díaz J, Roa K, Boulett A, Azócar L, Sánchez J. Reusable aminated lignin–based hydrogel biocomposite for effective dye adsorption in wastewater. International Journal of Biological Macromolecules. 2025;**304**:140842

[19] Zhou Y, Zhang Y, Wang X, Diao J, Zhang Y. Highly efficient adsorption of $Hg^{2+}$ from aqueous solutions by amino-functionalization alkali lignin. International Journal of Biological Macromolecules. 2022;**222**:3034-3044

[20] Ponomarev N, Pastushok O, Repo E, Doshi B, Sillanpää M. Lignin-based magnesium hydroxide nanocomposite. Synthesis and application for the removal of potentially toxic metals from aqueous solution. ACS Applied Nano Materials. 2019;**2**:5492-5503

[21] Shi X, Hong J, Li J, Kong S, Song G, Naik N, et al. Excellent selectivity and high capacity of As (V) removal by a novel lignin-based adsorbent doped with N element and modified with $Ca^{2+}$. International Journal of Biological Macromolecules. 2021;**172**:299-308

[22] Leudjo Taka A, Pillay K, Mbianda Yangkou X. Nanosponge cyclodextrin polyurethanes and their modification with nanomaterials for the removal of pollutants from waste water: A review. Carbohydrate Polymers. 2017;**159**:94-107

[23] Ji J, Yuan X, Zhao Y, Jiang L, Wang H. Mechanistic insights of removing pollutant in adsorption and advanced oxidation processes by sludge biochar. Journal of Hazardous Materials. 2022;**430**:128375

[24] Dayal L, Yadav K, Dey U, Das K, Kumari P, Raj D, et al. Recent advancement in microplastic removal process from wastewater - a critical review. Journal of Hazardous Materials Advances. 2024;**16**:100460

[25] Pan Y, Gao SH, Ge C, Gao Q, Huang S, Kang Y, et al. Removing microplastics from aquatic environments: A critical review. Environmental Science and Ecotechnology. 2023;**13**:100222

[26] Liu Q, Khor SM. Emerging absorption-based techniques for removing microplastics and nanoplastics from actual water bodies. TrAC - Trends in Analytical Chemistry. 2024;**170**:117465

[27] Găgeanu I, Carvalheiro F, Ekielski A, Duarte LC. Lignin utilization for the removal of microplastic particles from water. INMATEH - Agricultural Engineering. 2023;**71**:511-521

[28] Bahi A, Shao J, Mohseni M, Ko FK. Membranes based on electrospun lignin-zeolite composite nanofibers. Separation and Purification Technology. 2017;**187**:207-213

[29] Environment and Climate Change Canada. Releases of harmful substances

to water. In: Environment and Climate Change Canada. 2022. Available from: https://publications.gc.ca/site/archivee-archived.html?url=https://publications.gc.ca/collections/collection_2022/eccc/en4-144/En4-144-81-2022-eng.pdf [Accessed:15 May 2025]

[30] Liu C, Li Y, Gai X, Xiang Z, Jiang W, He S, et al. Advances in green materials derived from wood for detecting and removing mercury ions in water. Environmental Pollution. 2023;**335**:122351

[31] Xiao D, Ding W, Zhang J, Ge Y, Wu Z, Li Z. Fabrication of a versatile lignin-based nano-trap for heavy metal ion capture and bacterial inhibition. Chemical Engineering Journal. 2019;**358**:310-320

[32] Kong Y, Wang L, Ge Y, Su H, Li Z. Lignin xanthate resin–bentonite clay composite as a highly effective and low-cost adsorbent for the removal of doxycycline hydrochloride antibiotic and mercury ions in water. Journal of Hazardous Materials. 2019;**368**:33-41

[33] Zhou Y, Zhang J, Luo X, Lin X. Adsorption of Hg (II) in aqueous solutions using mercapto-functionalized alkali lignin. Journal of Applied Polymer Science. 2014;**131**:40979

[34] Zhang Z, Terrasson V, Guénin E. Lignin nanoparticles and their nanocomposites. Nanomaterials. 2021;**11**:1336

[35] Guo M, Jin T, Nghiem NP, Fan X, Qi PX, Jang CH, et al. Assessment of antioxidant and antimicrobial properties of lignin from corn Stover residue pretreated with low-moisture anhydrous ammonia and enzymatic hydrolysis process. Applied Biochemistry and Biotechnology. 2018;**184**:350-365

[36] Alzagameem A, Klein SE, Bergs M, Do XT, Korte I, Dohlen S, et al. Antimicrobial activity of lignin and lignin-derived cellulose and chitosan composites against selected pathogenic and spoilage microorganisms. Polymers. 2019;**11**:670

[37] Lizundia E, Sipponen MH, Greca LG, Balakshin M, Tardy BL, Rojas OJ, et al. Multifunctional lignin-based nanocomposites and nanohybrids. Green Chemistry. 2021;**23**:6698-6760

[38] Richter AP, Brown JS, Bharti B, Wang A, Gangwal S, Houck K, et al. An environmentally benign antimicrobial nanoparticle based on a silver-infused lignin core. Nature Nanotechnology. 2015;**10**:817-823

Chapter 5

# Lignin as a Sustainable Filler in Composites: Fabrications, Mechanical Properties, and Applications

*Jiawei Chen and Yu Chen*


## Abstract

This chapter reviews the recent progress of polymer composites that use lignin as reinforcing fillers, including the production methodologies and applications. Lignin is a naturally abundant and renewable bio-polymer. Due to its high stiffness, it usually serves as the structural component when integrated into polymer matrices. The addition of lignin as reinforcement of thermoplastic and elastomeric systems not only improves mechanical properties but also introduces various functional properties such as biocompatibility, antioxidant, and antimicrobial effects. These composites can be produced with a variety of morphologies and particle sizes, which allows them to be used in different applications such as smart materials, biomedical devices, structural components, and energy storage. The expanding interest in lignin-based composites reflects an increasing emphasis on the development of sustainable materials in the area of advanced composites.


**Keywords:** mechanical properties, composites, reinforcement, fillers, lignin

## 1. Introduction

In human history, natural materials that are derived from bioresources have long been used in a variety of ways due to their renewability, versatility, and availability [1]. Among different types of naturally derived materials, lignin is one of the most abundant natural polymers. In plant cell walls, lignin provides structural rigidity and microbial resistance [2, 3]. However, lignin is a side-product in the pulp and paper industry, and the majority of industrial "waste" lignin has been burned for energy recovery [4, 5]. Thus, its potential as a structural reinforcement has been substantially underestimated [2]. Studies have shown that lignin has an annual global production of approximately 70 million tons, while less than 2% of it is used in high-value applications such as carbon fiber precursors, bio-based adhesives, batteries, drug deliveries, water purification, and reinforcement for polymer composites [4, 5]. The main reason for this underutilization is primarily due to the heterogeneity of lignin's chemical structure, which results in significant challenges in its processing and integration into engineering materials.

Lignin has great potential as a filler in polymer composites since it has good compatibility with various types of polymers. Good compatibility will allow effective matrix integration [3, 6]. In addition, lignin has good thermal stability and chemical versatility. The characteristics provide opportunities for lignin to be modified or functionalized. As a result, lignin can be used as a low-cost alternative to conventional petroleum-based reinforcements [3, 6]. Nevertheless, the mechanical properties of lignin vary significantly based on the sources and the processing methods [7]. Based on its source, lignin can be classified as softwood lignin, hardwood lignin, or grass lignin. Each of these types of lignin exhibits different stability and chemical properties [8]. Typically, softwood lignin shows higher thermal stability due to its higher degree of crosslinking in the microstructure compared with other types. This compact and crosslinked structure leads to better suitability for processing methodologies that require high temperatures [3, 9]. On the contrary, hardwood lignin has a more linear structure, which makes it easier to be chemically modified [8]. Grass lignin has a wide range of structural variations and remains relatively unexplored in current research [8]. Beyond its sources, lignin can also be classified as sulfur-containing lignin and sulfur-free lignin, which is ultimately due to the difference in the pulping process. The most common one is the sulfur-containing lignin, such as kraft lignin and lignosulfonates. On the other hand, sulfur-free lignin, such as organosolv lignin, usually exhibits high purity and low ash content, but the process to extract sulfur-free lignin usually involves the use of organic solvents, which significantly increases the production cost and equipment deterioration [3, 5].

The mechanical properties of lignin have a large variation based on the extraction methods and sources. For example, the elastic modulus of lignin can vary from 1.2 to 14.8 GPa [10]. Regardless, the mechanical properties of lignin are usually sufficient to act as reinforcement for common thermoplastic-based polymers, such as polylactic acid (PLA), polypropylene (PP), polystyrene (PS), and polyethylene (PE). These polymers usually possess elastic moduli in the range of 0.7–3.4 GPa [11, 12]. Therefore, there is significant research potential in utilizing lignin as a filler in polymer composites.

Lignin-reinforced polymer composites can be produced with various fabrication methods, which offer different product morphologies, thermal properties, chemical properties, and mechanical properties [6]. Commonly seen fabrication methods include solution casting or spin coating, extrusion, 3D or 4D printing, and electrospinning. With these methods, one can obtain uniformly dispersed lignin in polymer matrices or have lignin modified to improve the interfacial adhesion for better dispersion. Compared to virgin polymers, the lignin-reinforced polymer composites exhibit improved tensile strength, elastic modulus, compressive strength, flame retardancy, etc. [13].

In this chapter, several fabrication techniques used to incorporate lignin into polymer composites have been compiled, and the applications of the composites have been summarized. Through these investigations, we aim to understand how lignin can be used as a sustainable alternative to conventional reinforcement fillers by discussing lignin's versatility in processing, compatibility with different polymer matrices, and its ability to enhance mechanical performance.

## 2. Composite fabrication and mechanical improvement

### 2.1 Extrusion

In polymer composite processing, extrusion is probably the most widely used technique due to its scalability [14–17]. Extrusion involves continuous feeding of a mixture of polymer pellets and lignin powders through an extruder, where high temperature and high shear force facilitate the dispersion of lignin particles into the melted polymer phase. The processing parameters include the choice of single-screw and twin-screw extruders, screw speed, temperature profiles, residence time, etc. These processing parameters can influence the dispersion quality, lignin orientation in the polymer matrix, and the overall mechanical properties of the resulting composites [6]. Literature has demonstrated that lignin can be blended with polyethylene (PE) [18], polyurethane (PU) [19], polypropylene (PP) [20], polylactic acid (PLA) [21], and other common thermoplastics in extrusion. To achieve a uniform dispersion of lignin in the polymer matrix, the processing temperature needs to exceed the melting temperature of the matrix polymer. This requirement could pose some challenges in the processing since the degradation temperature of lignin can be as low as 150°C, which is close to the melting temperature of a variety of polymers [18]. Despite that, lignin has been successfully incorporated into polymer matrices and typically yields enhanced elastic modulus and a reduced strain at break. This is due to the brittle nature of lignin compared to many virgin polymers [18, 20]. For instance, Gregorova et al. found that mixing 10 wt% lignin with PP improved the tensile strength from 26.9 ± 2.0 MPa to 32.4 ± 1.9 MPa and elastic modulus from 805 ± 30 to 940 ± 82 MPa, accompanied by reduced strain at failure from 456 ± 33 to 17 ± 2% [20]. Similarly, for recycled polystyrene (PS), Pérez-Guerrero et al. reported that 15 wt% lignin improved the storage elastic modulus of extruded recycled PS from 1629 MPa to approximately 1800 MPa [22]. These findings demonstrate lignin's potential as reinforcing fillers in the extrusion process.

To overcome the challenges and further improve the mechanical properties of the lignin-reinforced composites, compatibilizers and plasticizers can be used in the extrusion process [6]. This is the so-called reactive extrusion. In terms of chemical structure, the presence of phenolic and aliphatic hydroxyl groups in lignin's structure results in a reduced compatibility between lignin and those nonpolar matrices such as polybutylene adipate-co-terephthalate (PBAT) and PP [23]. Fortunately, chemical modification of lignin can be employed to address this challenge. Gordobil et al. used the acetylation process to modify kraft lignin and incorporate it in PLA using the extrusion process. They discovered that adding 10 wt% acetylated kraft lignin yielded an increase in elastic modulus, while adding 5 wt% acetylated kraft lignin led to an increase in strain at failure [17]. Liu et al. grafted lignin with vinyltrimethoxysilane (VTMS) to improve the dispersion and bonding with PBAT. They discovered that the reinforcement of 30 wt% of VTMS-grated lignin exhibited a 200% increase in tensile strength and a 151% increase in elastic modulus compared to pure PBAT [24]. Alternatively, the polymer matrix can also be modified to enhance compatibility. For example, PP can be grafted by maleic anhydride to be blended with 60 wt% of lignin with the twin-screw extrusion technique. The produced composites exhibited around 91% of the tensile strength and 336% of the elastic modulus of neat PP [25]. Furthermore, modifying both lignin and polymer matrix is an option to achieve better adhesion. As an example, Rojas-Lema et al. introduced dicumyl peroxide (DCP) to allow the formation of free radicals for both lignin and PE and compared the properties with PE/lignin composites without DCP. They discovered that the introduction of DCP increased strain at break while decreasing elastic modulus [23].

Other than compatibilizers, the addition of plasticizers can also improve the processability of lignin/polymer extrusion process [26]. Plasticizer can lower the glass-transition temperature of lignin and/or the polymer matrices, which reduces the viscosity and enhances the flow during the extrusion process. As a result, a more homogeneous melting can be produced. The typical plasticizers are polyethylene glycol (PEG), ε-caprolactone, ethylene carbonate, glycerol, vanillin, etc. [27, 28]. For example, Park et al. reinforced polycaprolactone (PCL) using kraft lignin, with ε-caprolactone as a plasticizer and polymeric diphenylmethane diisocyanate (pMDI) as a compatibilizer [21]. Compared with pure PCL, the plasticized composite with compatibilizer showed increased tensile strength and elastic modulus. Adding the plasticizer also increased the strain at break compared to the neat composite.

It is worth noting that compatibilizers and plasticizers may not always have a positive effect on extruded lignin-reinforced composites. In another study conducted by Park et al., the authors blended PLA with up to 30 wt% kraft lignin. ε-caprolactone and pMDI were used as the plasticizer and the compatibilizer, respectively [26]. As a comparison, without plasticizer or compatibilizer, the addition of kraft lignin led to a reduction in both tensile strength and strain at break of the PLA matrix. When plasticizer was added, the tensile strength and strain at break further reduced, while the elastic modulus showed no significant change. As a result, compatibilizer was further added into the blend, and increased tensile strength was demonstrated due to better interfacial adhesion and a smoother fracture surface. Nevertheless, the addition of both compatibilizer and plasticizer did not improve the tensile strength and strain at failure of the composites compared with that of virgin PLA. This study suggests that there is still considerable potential for research in optimizing the selection of compatibilizer and plasticizer.

## 2.2 3D/4D printing

The difference between 3D printing and 4D printing is the "4th dimension." In 3D printing, polymer or composite materials are deposited layer by layer to create the targeted geometries, while in 4D printing, a "4th dimension" of stimuli is involved to allow the 3D printed structure to change its shape or function after the formation of the geometry. In literature, lignin has been successfully incorporated into several 3D printing technologies, including Fused Deposition Modeling (FDM), Stereolithography (SLA), and Direct Ink Writing (DIW). For example, lignin has been blended with PLA or polyamide in FDM process to enhance mechanical properties or improve sustainability [16, 17]. In SLA, chemically modified lignin can be used as a reactive component in UV-curable resins to improve stiffness and crosslinking density. Lignin has also been used in DIW. For example, lignin-rich pastes or gels can be combined with other biomass constituents to create stable, shear-thinning inks [29]. With a similar idea as compared to reactive extrusion, chemical modification and functionalization of lignin can also be used in a 3D printer to enhance the compatibility and adhesion of lignin with various polymer matrices [29]. The most commonly used techniques are esterification, acetylation, and copolymerization. These modification techniques improve the dispersion of lignin within the polymer matrices. This subsection summarizes common 3D and 4D printing techniques to fabricate lignin-reinforced polymer composites. The role of lignin in enhancing the mechanical properties of the printed product is also discussed.

Another name for FDM is extrusion-based additive manufacturing, which means that FDM actually mirrors the principles of traditional extrusion but typically with a much smaller nozzle diameter. In preparation for 3D printing, lignin is usually first extruded with polymer pellets in order to produce filaments that can be used for 3D printing. The majority of recent research related to 3D printing lignin-reinforced composites focuses on filament-based FDM. For example, when lignin is incorporated in PLA for 3D printing, the mechanical properties often decrease as lignin content increases, which is primarily due to poor dispersion and agglomeration of lignin within the PLA matrix in the filament [30]. In addition, the amount of reduction in mechanical properties has been found to be related to the type of lignin as well [31]. However, studies have shown that the decrease in mechanical properties of lignin-reinforced PLA filaments for 3D printing can be partially resolved by the addition of plasticizers such as PEG or struktol® TR451. These plasticizers are able to enhance interfacial adhesion and yield comparable mechanical performance to that of virgin PLA [32]. Chemical modification can also be used to make further improvements. For example, Ding et al. modified lignin by copolymerizing it with 2-ethylhexylacrylate (EHA). The copolymerization technique led to an increase in strain at break from 2.98 ± 0.10% for 3D printed PLA to 21.7 ± 0.94%, with slightly reduced tensile strength and elastic modulus [33]. Recently, Anuar et al. demonstrated that epoxidized palm oil (EPO) is also an effective plasticizer for PLA/soda lignin composites. The printed composites showed an increase in tensile strength, elastic modulus, and strain at failure at different ratios of lignin and EPO [34]. Other than PLA, polyamide 12 (PA12) has also been fabricated into filaments for 3D printing with the reinforcement of lignin. For example, Zhang et al. modified hardwood organosolv lignin to introduce more phenolic hydroxyl groups, which enhanced adhesion between lignin and PA12 matrix. The composite specimens were printed with a nozzle temperature of 250°C. Addition of 10 wt% modified lignin in polyamide increased the tensile strength from 40.2 to 46.1 MPa, and elastic modulus from 1.48 to 1.73 GPa [35].

In SLA process, UV is used to cure the resin point by point on the surface. Lignin can be incorporated, although the concentrations are often very low compared to those of FDM [29]. For example, Zhang et al. used a solution blending method with acetone to mix low concentrations of kraft lignin (softwood) up to 1 wt% in the methacrylate resin and performed SLA using the produced composite resin. Post-curing was required due to the UV-blocking nature of lignin. After post-curing, the highest tensile strength of 49.6 MPa was obtained at 0.4 wt% lignin, and the highest elastic modulus of 2.39 GPa was obtained at 0.8 wt% lignin, compared to tensile strength of 30.3 MPa and elastic modulus of 1.74 GPa for post-cured resin without lignin [36]. To address this challenge, Sutton et al. modified organosolv lignin, which involved four types of treatments: methacrylic anhydride acylation, acrylic anhydride acylation, sodium borohydride reduction, and a combined treatment of reduction followed by methacrylic anhydride. Among all of these, lignin treated by both reduction and methacrylic anhydride demonstrated the best mechanical performance when blending the treated lignin with commercial SLA resin. This led to an increase in elastic modulus from around 275 MPa to around 625 MPa, tensile strength from around 38 MPa to around 65 MPa, and a reduction in strain at break [37].

In fact, DIW is another extrusion-based technique that builds the structure layer by layer by depositing viscoelastic, shear-thinning inks with a nozzle [38]. The major difference between FDM and DIW is that DIW typically operates at room temperature. However, to date, there have not been extensive studies conducted on lignin-reinforced composites produced by DIW, and the most extensively investigated bio-based ink for DIW is hydroxypropyl cellulose (HPC). It has been discovered that the incorporation of lignin in HPC can enhance the stability and alignment of the ink jet during shear-induced flow, and its oligomer may function as a chain stiffener to preserve the oriented structure. As a result, the mechanical performance may be enhanced [39]. In order to demonstrate the feasibility, Gleuwitz et al. produced shear-cast HPC films with organosolv lignin at an HPC/lignin ratio of 4:1 and successfully used the materials for DIW. However, cross-linking agents such as citric acid, unsaturated C18 fatty acid, and glycerol at different residual film moistures were necessary. The major discovery in their study is that, when printed in longitudinal direction aligned with tensile testing axis, the cross-linked films exhibited higher tensile strength and elastic modulus than the non-crosslinked films at a residual moisture of 6.4% [38]. Furthermore, Ebers and Laborie increased the lignin ratio to 50 wt% and printed composite monolayers with tensile strength of 18.5 MPa and elastic modulus of 995 MPa [40]. Recently, Yapa et al. used a combination of crosslinkers to improve elastic modulus from 316.2 ± 19.9 MPa to 375.6 ± 35.3 MPa and tensile strength from 3.7 ± 0.4 MPa to 4.2 ± 0.5 MPa for printing lignin/HPC composite by DIW [41]. Extending beyond room temperature DIW, Yang et al. introduced a frozen DIW technique to fabricate lignin/PVA composite structures with 15 wt% liquefied lignin. This approach yielded lignin/PVA gel ink extruded at a low temperature and freeze-cured to preserve its printed form. The fabricated structure was subsequently carbonized to produce carbon electrodes [42].

As discussed earlier in this subsection, the development of smart or stimuli-responsive lignin-reinforced materials for 4D printing is also a trend in the literature. 4D printing allows shape transformations triggered by external stimuli such as heat, light, or moisture [43]. For instance, Jin et al. incorporated up to 60 wt% enzymatically hydrolyzed lignin into epoxy soybean oil (ESO) and PEG by copolymerization and casted into film shape. The produced films have light-driven shape memory function, which is attractive to certain applications. They discovered that increasing lignin loading from 40 to 60 wt% yielded an increment in tensile strength for the shape memory polymer composite from 11.3 to 30.8 MPa. The shape fixation ratio was >98%, and the shape recovery ratio was >97% upon simulated solar stimulation [44]. This work suggests the potential of lignin as a filler in 4D printed structures. Later, Suo et al. successfully produced light-responsive lignin-reinforced shape memory polymer composites *via* FDM using PCL and thermoplastic polyurethane (TPU) as matrix. Enzyme-assisted lignin was used as a filler with a concentration up to 25 wt%. The elastic modulus of the printed composite increased with lignin content up to 20 wt%. The composites exhibited a shape recovery ratio of >95.06% under illumination [45].

## 2.3 Solution casting/spin coating

To obtain thin composite films with precisely controlled thickness, composition, and morphology, solution casting and spin coating are the two most widely utilized methods, especially for mixing lignin into polymer matrices at a micro-scale [46]. More specifically, solution casting involves dissolving polymer and lignin in a proper solvent and pouring it into a mold or onto a substrate to dry in a controlled environment to form a uniform film [47]. On the contrary, spin coating is to deposit a small amount of the solution onto a spinning substrate and form the thin film with centrifugal force after the solvent evaporates [48]. Both methodologies can be used with a wide range of polymer systems and allow the exploration of the effect of lignin on the mechanical properties of the composites.

Because of the dissolution process, solution casting and spin coating enable easier and more uniform dispersion of lignin within polymer matrices, as compared to melt-processing techniques [49]. Chemical modification of lignin can be performed so that it can be used as reinforcing fillers for elastomeric matrices. Shorey et al. discovered that incorporating 5 wt% silylated lignin in rubber to produce solution cast film improved tensile strength from 23.6 to ~34 MPa [49]. On the contrary, lignin can be used as reinforcement for thermoplastic matrices with or without chemical modifications. Makri et al. attempted to incorporate nanosized soda lignin in PLA by solution casting, then melt-mixing for up to 5 wt%. The highest tensile strength of 23.86 MPa was achieved at 2.5 wt% lignin with elastic modulus of 1300 ± 90 MPa, compared to virgin PLA with tensile strength of 8.41 MPa and elastic modulus of 638 MPa [50].

As a summary, due to the simple setup and scalability of solution casting and spin coating, they have become one of the most effective techniques for incorporating lignin into polymer matrices and exploring the effect of lignin on the properties of these composites.

### 2.4 Molding

Molding is another melt-processing technique that is widely used in polymer and composite fabrication. The two most commonly seen molding methods are injection molding and compression molding, and both have been reported in the literature for making lignin-reinforced polymer composites. In injection molding, lignin is mixed with thermoplastic polymer matrices and injected into a mold under high pressure. In compression molding, the lignin and polymer matrices are placed in a heated mold, and high pressure is obtained by compression. The most important advantage of molding compared to other fabrication techniques is that molding can produce parts with a great surface finish and a high filler loading [51].

Several existing studies have demonstrated the success of using molding techniques to fabricate lignin-reinforced polymer composites. For example, Intapun et al. compression-molded Klason lignin-reinforced natural rubber composites. The produced composites had an increased tensile strength and elastic modulus compared to that of pristine natural rubber [52]. In another study, Mo et al. reported that reinforcing polyvinyl alcohol (PVA) with alkaline lignin obtained compression-molded composites with the highest tensile strength of 71 ± 8 MPa and elastic modulus of 1337 ± 113, compared to PVA with tensile strength of 55 ± 5 MPa and elastic modulus of 345 ± 16 MPa [53]. For injection molding, Abdelwahab et al. developed PP composites containing 20 wt% lignin and 5 wt% EBGMA as a compatibilizer. The addition of the compatibilizer increased strain at failure from 16 to 162%. The impact strength of the composites was also found to be increased [54].

### 2.5 Electrospinning

Another technique to produce lignin-reinforced polymer composites is electrospinning. It is a versatile fabrication methodology to produce composite fibers down to the micro- and nano-scale. In electrospinning, a charged jet of the lignin-containing polymer solution or melt is ejected by the high electric field from a nozzle toward a grounded collector or an electrode with opposite polarity. During the flight, instability occurs, which significantly increases the travel distance of the jet. As a result, the solvent evaporates, or the melt solidifies to form solid fibers at the collector as nonwoven mats. The morphology of the produced mats can be tuned with the change of processing parameters such as solution concentration, voltage, nozzle-to-collector distance, etc. In the literature, the majority of the recent studies have focused on the electrospinning of lignin-reinforced nanofibers for low-cost precursors of carbon fibers [55]. The application of these electrospun composites has been targeted at biomedical applications due to their nano-/micro-scale dimensions and biocompatibility.

For example, Kai et al. demonstrated that blending 50% lignin copolymerized with poly(methyl methacrylate) (PMMA) into PCL enhanced the elastic modulus from 7.6 ± 0.3 MPa to 44.4 ± 12.2 MPa [56]. In a subsequent study, they also demonstrated that incorporating lignin copolymerized with PCL and poly(3-hydroxybutyrate) (PHB) improved the tensile strength of PHB from 1.81 ± 0.04 MPa to 3.13 ± 0.10 MPa, with elastic modulus increasing from 80.9 ± 2.1 MPa to 86.1 ± 1.0 MPa, and strain at failure increasing from 15 ± 1% to 55 ± 5% [57]. Similarly, Norouzi et al. reported inclusion of 10% alkaline lignin in electrospinning poly(lactic-co-glycolic acid) (PLGA) increased strain at failure from 35.89 ± 1.82% to 87.08 ± 2.68% [58]. As can be seen from these studies, the electrospinning process allows for the production of lignin-reinforced polymer composite nanofibers with enhanced mechanical properties for tissue engineering, wound dressings, and biomedical strain sensors [58].

## 3. Application of lignin composites

In this subsection, some of the typical applications of lignin-based composites are explored, including smart materials, biomedical fields, structural materials, and energy materials [59]. From these applications, it can be seen that lignin not only enhances the mechanical properties and sustainability of the composites but also provides functionalities in various fields.

### 3.1 Smart materials

The main constituent of lignin's structure is phenylpropane units, which makes lignin a photothermal agent for absorbing light at different wavelengths [60]. When fabricating smart materials made from elastomeric composites with light-driven shape memory properties, lignin's photothermal capability allows it to convert light energy into heat energy as a filler to activate the shape memory effect [61]. On the other hand, lignin can also be used to reinforce these smart polymers to achieve high-performance stimuli-responsive composite materials and provide enhanced sustainability [44]. The improved mechanical properties contributed by the presence of lignin allow the stimuli-responsive composite materials to exhibit better structural integrity, especially when cycling motion is required for the application. Lignin-reinforced composites are not restricted to shape memory applications. Lignin can be combined with polyaniline to form a solar interfacial evaporator with balsa wood as a substrate [62]. Solar interfacial evaporator absorbs light and centralizes heat at the air-water interface to generate water vapor in a sustainable manner [62]. Lignin can also be combined with chitosan to form films for smart windows, which can reduce heating energy by increasing indoor temperature through photothermal activity [63].

### 3.2 Biomedical applications

In terms of biomedical applications, lignin exhibits antioxidant, antimicrobial, and biocompatible properties. These properties make lignin suitable for tissue engineering, drug delivery, and wound dressings [59]. For instance, lignin can be blended with PHB or PCL to create scaffolds for tissue cell culture [56, 57]. In tissue engineering, lignin can induce the breakage of bacteria and demonstrate good cell affinity to promote cell culturing [64]. It can also be used for strengthening hydrogels and providing antibacterial properties for wound dressing applications [59]. Lignin-reinforced hydrogel also manifests self-healing properties through forming ionic bonds when two cut surfaces are in contact, thereby lengthening its service time [65].

### 3.3 Construction

In other studies, lignin-reinforced composites have been developed for protective coatings for construction materials, which offer water resistance and improved thermal and mechanical properties [66]. When lignin is mixed with coating materials such as epoxy, it can provide UV protection due to its UV-blocking property and prevent corrosion or wear for steel [67]. On account of its antimicrobial properties, lignin protects the construction material by slowing down environmental degradation. Other than protective film or coating, lignin can also be used as a concrete admixture and insulation in construction. Mixing lignin in concrete can improve water resistance, antibacterial capabilities, and mechanical properties of concrete [68]. Adding pulp waste to concrete has also been demonstrated to increase the thermal insulation of concrete [68]. Overall, lignin provides anti-corrosion, UV-blocking, wear-resistance, water resistance, and antibacterial features in construction, along with improved mechanical properties.

### 3.4 Adhesives

Adhesive is another application for lignin-reinforced composites. Conventional adhesive materials, such as formaldehyde or phenol, are petroleum-derived and pose negative effects on the environment. In addition, formaldehyde is toxic and has a relatively high cost, while most formaldehyde-free, eco-friendly adhesives have low water resistance or undesired durability [69]. Adding lignin to these adhesive materials, like polyethylenimine or tannin, yields improved bonding in a wet environment and accelerated curing [69]. Lignin can improve the mechanical properties of the adhesive material, but modification or compatibilizers may be required to achieve the required crosslinking [70].

### 3.5 Energy applications

As mentioned in the previous subsection, lignin-reinforced composites have been produced as precursors for carbon fibers and carbon-based electrodes due to lignin's high carbon content. The produced electrodes obtain improved capacity in battery applications [71]. Lignin-reinforced composites can also be used in supercapacitors, binders, electrolytes, and separators [71]. Thus, lignin offers a renewable alternative to petroleum-derived materials for energy storage and conversion technologies [42].

Many other applications, such as food packaging or cosmetics, have not been elaborated on in this subsection [72]. In a nutshell, all these advancements have emphasized lignin's potential as a sustainable and multifunctional filler across a broad range of high-performance composite applications.

## 4. Conclusion

In this chapter, lignin's role as a reinforcement filler for polymer composites has been discussed. Lignin can be incorporated into various polymer matrices, including thermoplastic and elastomers. The lignin-reinforced composites have improved mechanical properties, and thus, they have been demonstrated to be used in a wide range of applications. A list of works for lignin-reinforced composites is summarized below (**Table 1**). Overall, incorporating lignin in polymeric matrices provides a cost-effective solution for producing high-performance composites.

| Polymer | Lignin | Lignin Ratio (wt%) | Fabrication Method | Modification | Mechanical Properties | Ref |
|---|---|---|---|---|---|---|
| PS | Kraft Lignin | 15 | Extrusion | Acetylated, Malleated Lignin | Unmodified, acetylated, and malleated lignin increased storage modulus of recycled PS from 1600 to 1818 MPa, 1759 MPa, and 1802 MPa, respectively | [22] |
| PE | Kraft Lignin | 60 | Extrusion | Ethylene-glycidylmethacrylate copolymer as compatibilizer | Elastic modulus increases, tensile strength, and strain at break decreased with lignin content | [18] |
| Bio-PE | Kraft Lignin | 20 | Extrusion | DCP | Elastic modulus increased after lignin was added; strain at break was improved after adding DCP compared to bio-based PE/lignin | [23] |
| PP | Kraft Lignin | 60 | Extrusion | maleic anhydride grafted PP | tensile strength reduced to 91% but elastic modulus increased to 336% after lignin was added | [25] |
| PBAT | Kraft Lignin | 30 | Extrusion | VTMS-grafted lignin | 200% increase in tensile strength and a 151% increase in elastic modulus | [24] |
| PLA | Kraft Lignin | 30 | Extrusion | ε-caprolactone as plasticizer and pMDI as compatibilizer | Tensile strength increased after adding pMDI to plasticized lignin/PLA, but mechanical properties reduced compared to PLA | [26] |
| PLA | Kraft Lignin | 10 | Extrusion | Acetylated lignin | Tensile strength increased at 10 wt%, and strain at failure increased at 5 wt% | [17] |
| PCL | Kraft Lignin | 30 with pMDI; 50 without pMDI | Extrusion | ε-caprolactone as plasticizer and pMDI as compatibilizer | Tensile strength increased by 20.2%, and elastic modulus improved by 91.8% | [21] |
| PA12 | Organosolv Lignin | 10 | FDM | Demethylation | Tensile strength increased by 14.7%, and elastic modulus increased by 16.9% | [35] |
| PLA | Kraft Lignin | 10 | FDM | Copolymerization with 2-ethylhexylacrylate (EHA) | Strain at break increased by 628% | [33] |

| Polymer | Lignin | Lignin Ratio (wt%) | Fabrication Method | Modification | Mechanical Properties | Ref |
|---|---|---|---|---|---|---|
| PLA | Soda Lignin | 1–5 phr | FDM | EPO | Tensile strength, elastic modulus, and strain at break increased at various ratios of lignin/EPO | [34] |
| Resin | Kraft Lignin | 1 | SLA | N/A | Tensile strength increased by 46–64%, and elastic modulus increased by 13–37% | [36] |
| Resin | Organosolv Lignin | 5 | SLA | methacrylic anhydride acylation, acrylic anhydride acylation, sodium borohydride reduction, and reduction followed by methacrylic anhydride | Elastic modulus increased from ~275 to ~625 MPa, and tensile strength increased from ~38 to ~65 MPa | [37] |
| HPC | Organosolv Lignin | 20 | DIW | citric acid, fatty acid, and glycerol | Tensile strength and elastic modulus increased after crosslinkers were added at print orientation of 0° and moisture content of 6.4% | [38] |
| HPC | Organosolv Lignin | 50 | DIW | citric acid and borate ions | Tensile strength of 18.5 MPa and elastic modulus of 995 MPa | [40] |
| HPC | Organosolv Lignin | 50 | DIW | SpeedCure BPO, SpeedCure 938, and Tris | Tensile strength of 4.2 MPa and elastic modulus of 375.6 MPa | [41] |
| PEG and ESO | Enzymatically hydrolyzed lignin | 50 | Cast | N/A | Tensile strength increased from 11.3 to 30.8 MPa as the lignin ratio increased from 40 to 60 wt% | [44] |
| PCL and TPU | Enzyme-assisted Lignin | 25 | FDM/4D Print | N/A | Elastic modulus increased by almost 50% after addition of 20 wt% lignin | [45] |
| Rubber | Kraft Lignin | 5 | Solution Cast | silylated Lignin | Tensile strength increased by 44% | [49] |
| PLA | Soda Lignin | 5 | Solution Cast | N/A | Tensile strength increased by ~184%, and elastic modulus increased by ~115% | [50] |
| PVA | Alkaline Lignin | 5 phr | Compression Molding | Ethanol extraction | Tensile strength increased by 29%, and elastic modulus increased by 288% | [53] |

| Polymer | Lignin | Lignin Ratio (wt%) | Fabrication Method | Modification | Mechanical Properties | Ref |
|---|---|---|---|---|---|---|
| PP | Lignin | 20–30 | Injection Molding | EBGMA and/or MAPP | Impact strength increased by 28%, and strain at failure increased by ~913% after adding 5 wt.% EBGMA | [54] |
| Rubber | Klason Lignin | 20 phr | Compression Molding | N/A | Improved strength and modulus | [52] |
| PHB | Alkaline Lignin | 5 | Electrospinning | Copolymerized with PCL and PHB | Tensile strength improved by ~73%, elastic modulus increased by ~6%, and strain at failure increased by ~267% | [57] |
| PLGA | Alkaline Lignin | 10 | Electrospinning | N/A | Strain at failure increased by ~143% | [58] |

**Table 1.**
*Summary of mechanical properties of polymer composites with lignin as a filler.*

## Conflict of interest

The authors declare no conflict of interest.

## Nomenclature

| | |
|---|---|
| DIW | direct ink writing |
| EPO | epoxidized palm oil |
| ESO | epoxy soybean oil |
| FDM | fused deposition modeling |
| HPC | hydroxypropyl cellulose |
| PA | polyamide |
| PBAT | adipate-co-terephthalate |
| PCL | polycaprolactone |
| PE | polyethylene |
| PEG | polyethylene glycol |
| PHB | poly(3-hydroxybutyrate) |
| PLA | polylactic acid |
| PMMA | poly(methyl methacrylate) |
| PVA | polyvinyl alcohol |
| PP | polypropylene |
| PS | polystyrene |

## Author details

Jiawei Chen* and Yu Chen
University of Alberta, Edmonton, Alberta, Canada

*Address all correspondence to: jiawei9@ualberta.ca

## References

[1] Olatunji O. Natural Polymers. Cham, Heidelberg, New York, Dordrecht, London, Switzerland: Springer International Publishing; 2016

[2] Melro E, Filipe A, Sousa D, Medronho B, Romano A. Revisiting lignin: A tour through its structural features, characterization methods and applications. New Journal of Chemistry. 2021;**45**:6986-7013

[3] Doherty WOS, Mousavioun P, Fellows CM. Value-adding to cellulosic ethanol: Lignin polymers. Industrial Crops and Products. 2011;**33**:259-276

[4] Bajwa DS, Pourhashem G, Ullah AH, Bajwa SG. A concise review of current lignin production, applications, products and their environmental impact. Industrial Crops and Products. 2019;**139**:111526

[5] Lora JH, Glasser WG. Recent industrial applications of lignin: A sustainable alternative to nonrenewable materials. Journal of Polymers and the Environment. 2002;**10**:39-48

[6] Thakur VK, Thakur MK, Raghavan P, Kessler MR. Progress in green polymer composites from lignin for multifunctional applications: A review. ACS Sustainable Chemistry & Engineering. 2014;**2**:1072-1092

[7] Huang D, Li R, Xu P, Li T, Deng R, Chen S, et al. The cornerstone of realizing lignin value-addition: Exploiting the native structure and properties of lignin by extraction methods. Chemical Engineering Journal. 2020;**402**:126237

[8] Lu Y-C, Lu Y, Fan X. Structure and characteristics of lignin. In: Lignin: Biosynthesis and Transformation for Industrial Applications. Switzerland AG: Springer Nature; 2020. pp. 17-75

[9] Kadla JF, Kubo S, Venditti RA, Gilbert RD, Compere AL, Griffith W. Lignin-based carbon fibers for composite fiber applications. Carbon N Y. 2002;**40**:2913-2920

[10] Schwaighofer M, Zelaya-Lainez L, Königsberger M, Lukacevic M, Serna-Loaiza S, Harasek M, et al. Characterization of mechanical properties of five hot-pressed lignins extracted from different feedstocks by microscopy-aided nanoindentation. Materials and Design. 2023;**227**:111765

[11] Graziano A, Jaffer S, Sain M. Review on modification strategies of polyethylene/polypropylene immiscible thermoplastic polymer blends for enhancing their mechanical behavior. Journal of Elastomers & Plastics. 2019;**51**:291-336

[12] Ranakoti L, Gangil B, Mishra SK, Singh T, Sharma S, Ilyas RA, et al. Critical review on polylactic acid: Properties, structure, processing, biocomposites, and nanocomposites. Materials. 2022;**15**:4312

[13] Agustiany EA, Rasyidur Ridho M, Rahmi DNM, Madyaratri EW, Falah F, Lubis MAR, et al. Recent developments in lignin modification and its application in lignin-based green composites: A review. Polymer Composites. 2022;**43**:4848-4865

[14] Anwer MAS, Naguib HE, Celzard A, Fierro V. Comparison of the thermal, dynamic mechanical and morphological properties of PLA-lignin & PLA-tannin particulate green composites. Composites Part B: Engineering. 2015;**82**:92-99

[15] Kumar Singla R, Maiti SN, Ghosh AK. Crystallization, morphological, and mechanical response of poly (lactic acid)/ lignin-based biodegradable composites. Polymer - Plastics Technology and Engineering. 2016;**55**:475-485

[16] Gordobil O, Egüés I, Llano-Ponte R, Labidi J. Physicochemical properties of PLA lignin blends. Polymer Degradation and Stability. 2014;**108**:330-338

[17] Gordobil O, Delucis R, Egüés I, Labidi J. Kraft lignin as filler in PLA to improve ductility and thermal properties. Industrial Crops and Products. 2015;**72**:46-53

[18] Tang Y, Jean M, Pourebrahimi S, Rodrigue D, Ye Z. Influence of lignin structure change during extrusion on properties and recycling of lignin-polyethylene thermoplastic composites. Canadian Journal of Chemical Engineering. 2021;**99**:S27-S38

[19] Zhou X, Ren Z, Sun H, Bi H, Gu T, Xu M. 3D printing with high content of lignin enabled by introducing polyurethane. International Journal of Biological Macromolecules. 2022;**221**:1209-1217

[20] Gregorova A, Košíková B, Staško A. Radical scavenging capacity of lignin and its effect on processing stabilization of virgin and recycled polypropylene. Journal of Applied Polymer Science. 2007;**106**:1626-1631

[21] Park C-W, Youe W-J, Namgung H-W, Han S-Y, Seo P-N, Chae H-M, et al. Effect of lignocellulose nanofibril and polymeric methylene diphenyl diisocyanate addition on plasticized lignin/polycaprolactone composites. BioResources. 2018;**13**:6802-6817

[22] Pérez-Guerrero P, Lisperguer J, Navarrete J, Rodrigue D. Effect of modified Eucalyptus nitens lignin on the morphology and thermo-mechanical properties of recycled polystyrene. BioResources. 2014;**9**:6514-6526

[23] Rojas-Lema S, Ivorra-Martinez J, Lascano D, Garcia-Garcia D, Balart R. Improved performance of environmentally friendly blends of biobased polyethylene and Kraft lignin compatibilized by reactive extrusion with dicumyl peroxide. Macromolecular Materials and Engineering. 2021;**306**:2100196

[24] Liu Y, Liu S, Liu Z, Lei Y, Jiang S, Zhang K, et al. Enhanced mechanical and biodegradable properties of PBAT/ lignin composites via silane grafting and reactive extrusion. Composites Part B: Engineering. 2021;**220**:108980

[25] Wang R, You X, Qi S, Tian R, Zhang H. Enhancing mechanical performance of high-lignin-filled polypropylene via reactive extrusion. Polymers (Basel). 2024;**16**:520

[26] Park C-W, Youe W-J, Kim S-J, Han S-Y, Park J-S, Lee E-A, et al. Effect of lignin plasticization on physico-mechanical properties of lignin/poly (lactic acid) composites. Polymers (Basel). 2019;**11**:2089

[27] Bouajila J, Dole P, Joly C, Limare A. Some laws of a lignin plasticization. Journal of Applied Polymer Science. 2006;**102**:1445-1451

[28] Kun D, Pukánszky B. Polymer/ lignin blends: Interactions, properties, applications. European Polymer Journal. 2017;**93**:618-641

[29] Wu T, Sugiarto S, Yang R, Sathasivam T, Weerasinghe UA, Chee PL, et al. From 3D to 4D printing of lignin towards green materials and sustainable manufacturing. Materials Horizons. 2025;**12**:2789-2819

[30] Zaidi SAS, Kwan CE, Mohan D, Harun S, Luthfi AAI, Sajab MS. Evaluating the stability of PLA-lignin filament produced by bench-top extruder for sustainable 3D printing. Materials. 2023;**16**:1793

[31] Mimini V, Sykacek E, Syed Hashim SNA, Holzweber J, Hettegger H, Fackler K, et al. Compatibility of Kraft lignin, organosolv lignin and lignosulfonate with PLA in 3D printing. Journal of Wood Chemistry and Technology. 2019;**39**:14-30

[32] Wasti S, Triggs E, Farag R, Auad M, Adhikari S, Bajwa D, et al. Influence of plasticizers on thermal and mechanical properties of biocomposite filaments made from lignin and polylactic acid for 3D printing. Composites Part B: Engineering. 2021;**205**:108483

[33] Ding R, Duan Z, Sun Y, Yuan Q, Tien TT, Zúniga MG, et al. Enhancement of 3D printability and mechanical properties of polylactic acid/lignin biocomposites via interface engineering. Industrial Crops and Products. 2023;**194**:116286

[34] Anuar H, Abd Rahman NA, Manshor MR, Alli YA, Alimi OA, Alif F, et al. Novel soda lignin/PLA/EPO biocomposite: A promising and sustainable material for 3D printing filament. Materials Today Communications. 2023;**35**:106093

[35] Zhang S, Meng X, Bhagia S, Ji A, Dean Smith M, Wang Y-y, et al. 3D printed lignin/polymer composite with enhanced mechanical and anti-thermal-aging performance. Chemical Engineering Journal. Feb 2024;**481**:148449. DOI: 10.1016/j.cej.2023.148449

[36] Zhang S, Li M, Hao N, Ragauskas AJ. Stereolithography 3D printing of lignin-reinforced composites with enhanced mechanical properties. ACS Omega. 2019;**4**:20197-20204

[37] Sutton JT, Rajan K, Harper DP, Chmely SC. Improving UV curing in organosolv lignin-containing photopolymers for stereolithography by reduction and acylation. Polymers (Basel). 2021;**13**:3473

[38] Gleuwitz FR, Sivasankarapillai G, Siqueira G, Friedrich C, Laborie M-PG. Lignin in bio-based liquid crystalline network material with potential for direct ink writing. ACS Applied Bio Materials. 2020;**3**:6049-6058

[39] Gleuwitz FR, Friedrich C, Laborie M-PG. Lignin-assisted stabilization of an oriented liquid crystalline cellulosic mesophase, part A: Observation of microstructural and mechanical behavior. Biomacromolecules. 2020;**21**:1069-1077

[40] Ebers L-S, Laborie M-P. Direct ink writing of fully bio-based liquid crystalline lignin/hydroxypropyl cellulose aqueous inks: Optimization of formulations and printing parameters. ACS Applied Bio Materials. 2020;**3**:6897-6907

[41] Yapa M-T, Lalevée J, Laborie M-P. Towards photocrosslinkable lyotropic blends of organosolv lignin and hydroxypropyl cellulose for 3D printing by direct ink writing. Polymers (Basel). 2024;**16**:2869

[42] Yang Z, Li J, Jiang C, Li B, Guo Y, Wang Y, et al. Technical and experimental studies on the preparation of lignin/PVA specially shaped carbon electrodes based on frozen DIW technology. Diamond and Related Materials. 2025;**155**:112361

[43] Ren Z, Ding K, Zhou X, Ji T, Sun H, Chi X, et al. 4D printing light-driven

actuator with lignin photothermal conversion module. International Journal of Biological Macromolecules. 2023;**253**:126562

[44] Jin X, Liu X, Li X, Du L, Su L, Ma Y, et al. High lignin, light-driven shape memory polymers with excellent mechanical performance. International Journal of Biological Macromolecules. 2022;**219**:44-52

[45] Suo F, Bai X, Liu Y, Xu M, Gu T, Cao L, et al. Development of lignin-based 3D-printable light responsive shape memory materials: Design of optically controlled devices. International Journal of Biological Macromolecules. 2024;**277**:132943

[46] Butt MA. Thin-film coating methods: A successful marriage of high-quality and cost-effectiveness—A brief exploration. Coatings. 2022;**12**:1115

[47] Chen L, Tang C, Ning N, Wang C, Fu Q, Zhang Q. Preparation and properties of chitosan/lignin composite films. Chinese Journal of Polymer Science. 2009;**27**:739-746

[48] Notley SM, Norgren M. Surface energy and wettability of spin-coated thin films of lignin isolated from wood. Langmuir. 2010;**26**:5484-5490

[49] Shorey R, Gupta A, Mekonnen TH. Hydrophobic modification of lignin for rubber composites. Industrial Crops and Products. 2021;**174**:114189

[50] Makri SP, Xanthopoulou E, Klonos PA, Grigoropoulos A, Kyritsis A, Tsachouridis K, et al. Effect of micro-and nano-lignin on the thermal, mechanical, and antioxidant properties of biobased PLA–lignin composite films. Polymers (Basel). 2022;**14**:5274

[51] Fu H, Xu H, Liu Y, Yang Z, Kormakov S, Wu D, et al. Overview of injection molding technology for processing polymers and their composites. ES Materials & Manufacturing. 2020;**8**:3-23

[52] Intapun J, Rungruang T, Suchat S, Cherdchim B, Hiziroglu S. The characteristics of natural rubber composites with Klason lignin as a green reinforcing filler: Thermal stability, mechanical and dynamical properties. Polymers (Basel). 2021;**13**:1109

[53] Mo J, Lei J, Wang H, Kang Q, Liu W, Qiu X. Melt-processable polyvinyl alcohol/lignin composites with improved strength via synergistic plasticization of lignin. International Journal of Biological Macromolecules. 2024;**267**:131726

[54] Abdelwahab MA, Misra M, Mohanty AK. Injection molded biocomposites from polypropylene and lignin: Effect of compatibilizers on interfacial adhesion and performance. Industrial Crops and Products. 2019;**132**:497-510

[55] Svinterikos E, Zuburtikudis I, Al-Marzouqi M. Electrospun lignin-derived carbon micro-and nanofibers: A review on precursors, properties, and applications. ACS Sustainable Chemistry & Engineering. 2020;**8**:13868-13893

[56] Kai D, Jiang S, Low ZW, Loh XJ. Engineering highly stretchable lignin-based electrospun nanofibers for potential biomedical applications. Journal of Materials Chemistry B. 2015;**3**:6194-6204

[57] Kai D, Chong HM, Chow LP, Jiang L, Lin Q, Zhang K, et al. Strong and biocompatible lignin/poly (3-hydroxybutyrate) composite nanofibers. Composites Science and Technology. 2018;**158**:26-33

[58] Norouzi M, Rafienia M, Hosseini S. Characterization and biological

evaluation of new PLGA/fibrin/lignin biocomposite electrospun scaffolds. Physica Scripta. 2023;**98**:095506

[59] Liu R, Dai L, Xu C, Wang K, Zheng C, Si C. Lignin-based micro-and nanomaterials and their composites in biomedical applications. ChemSusChem. 2020;**13**:4266-4283

[60] Guan Y, Liu H, Han Y, Zhou Y, Yan L, Dai L, et al. Unlocking the photothermal conversion capacity of lignin and lignin-derived materials. Chemical Engineering Journal. 2024;**497**:154985

[61] Li J, Liu W, Qiu X, Zhao X, Chen Z, Yan M, et al. Lignin: A sustainable photothermal block for smart elastomers. Green Chemistry. 2022;**24**:823-836

[62] Lu H, Li Q, Liu W, Li Z. Enhanced solar interfacial evaporation through lignin-polyaniline composite coatings on balsa wood substrates. International Journal of Biological Macromolecules. 2024;**281**:136382

[63] Liu J, Moreno A, Chang J, Morsali M, Yuan J, Sipponen MH. Fully biobased photothermal films and coatings for indoor ultraviolet radiation and heat management. ACS Applied Materials & Interfaces. 2022;**14**:12693-12702

[64] Liu Y, Wang X, Wu Q, Pei W, Teo MJ, Chen ZS, et al. Application of lignin and lignin-based composites in different tissue engineering fields. International Journal of Biological Macromolecules. 2022;**222**:994-1006

[65] Zhang Y, Yuan B, Zhang Y, Cao Q, Yang C, Li Y, et al. Biomimetic lignin/poly (ionic liquids) composite hydrogel dressing with excellent mechanical strength, self-healing properties, and reusability. Chemical Engineering Journal. 2020;**400**:125984

[66] Fazeli M, Mukherjee S, Baniasadi H, Abidnejad R, Mujtaba M, Lipponen J, et al. Lignin beyond the status quo: Recent and emerging composite applications. Green Chemistry. 2024;**26**:593-630

[67] Wang D, Claesson P, Zhang F, Pan J, An R, Shi Y. Recent findings on lignin-based wear and corrosion resistance coatings. Corrosion Reviews. 2025;**43**:157-174

[68] Patel R, Babaei-Ghazvini A, Dunlop MJ, Acharya B. Biomaterials-based concrete composites: A review on biochar, cellulose and lignin. Carbon Capture Science & Technology. 2024;**12**:100232

[69] Huang C, Peng Z, Li J, Li X, Jiang X, Dong Y. Unlocking the role of lignin for preparing the lignin-based wood adhesive: A review. Industrial Crops and Products. 2022;**187**:115388

[70] Ridho MR, Agustiany EA, Rahmi Dn M, Madyaratri EW, Ghozali M, Restu WK, et al. Lignin as green filler in polymer composites: Development methods, characteristics, and potential applications. Advances in Materials Science and Engineering. 2022;**2022**:1363481

[71] Beaucamp A, Muddasar M, Amiinu IS, Leite MM, Culebras M, Latha K, et al. Lignin for energy applications–state of the art, life cycle, technoeconomic analysis and future trends. Green Chemistry. 2022;**24**:8193-8226

[72] Priyadarshi R, Ghosh T, Purohit SD, Prasannavenkadesan V, Rhim J-W. Lignin as a sustainable and functional material for active food packaging applications: A review. Journal of Cleaner Production. 2024;**469**:143151

Chapter 6

# Application of Lignin in Petroleum Processes

*Abdelaziz El-hoshoudy*


## Abstract

Lignin, a natural and abundant biopolymer, has gained attention in petroleum processes, particularly in enhanced oil recovery (EOR). Its unique amphiphilic structure allows it to act as a surfactant, reducing interfacial tension (IFT) between oil and water, thus improving oil displacement efficiency. Modified lignin derivatives enhance viscosity, providing better mobility control and improving sweep efficiency during polymer flooding. Additionally, lignin-based nanoparticles can alter rock wettability, shifting it to a more water-wet state, and facilitating oil release. Its biodegradability and cost-effectiveness make it a promising alternative to synthetic EOR agents. Recent studies explore lignin-polymer composites and functionalized lignin in reservoir conditions, highlighting its potential to improve oil recovery while minimizing environmental impact.


**Keywords:** enhanced oil recovery, lignin, IFT reduction, polymer viscosity, biopolymers

## 1. Introduction

The shift toward renewable resources has become increasingly vital in the development of polymeric materials, particularly as substitutes for petroleum-derived plastics [1]. This movement is largely motivated by environmental concerns associated with fossil fuels and the search for sustainable, biodegradable alternatives such as lignin and its derivatives [2]. Simultaneously, the petroleum industry faces the dual challenge of rising global energy demand and declining output from aging oil fields. To address this, enhanced oil recovery (EOR) technologies have emerged, targeting the significant volumes of crude oil that remain trapped in porous reservoir rocks after conventional primary and secondary recovery methods that typically extract only about one-third of the original oil in place [3]. Among the chemical EOR methods, surfactant flooding has proven particularly effective. Surfactants significantly reduce the interfacial tension (IFT) between oil and water and modify the wettability of rock surfaces [4]. These changes facilitate the mobilization of residual oil by increasing the capillary number and promoting emulsification. The mechanisms of wetting behavior on micro-structured surfaces are often explained by the Wenzel and Cassie-Baxter models, which describe how liquid interfaces interact with rough surfaces either by complete surface wetting or by partial wetting with trapped air pockets. Traditional surfactants used in EOR are primarily anionic, including petroleum sulfonates and lignin sulfonates [5, 6]. However, their effectiveness is compromised in saline environments, particularly when divalent ions cause precipitation. To mitigate these limitations, recent research has focused on formulations that blend anionic and nonionic surfactants. Nonionic surfactants are especially valued for their cost efficiency, environmental compatibility, tolerance to brine salinity, and low adsorption on reservoir rocks [7].

Polymers are also a cornerstone of chemical EOR, enhancing fluid viscosity and sweep efficiency. Hydrolyzed polyacrylamide (HPAM) is commonly used for its low cost and water solubility [8]. However, it is prone to degradation under high-temperature and high-salinity conditions, resulting in increased injection volumes and higher operational costs [9]. In contrast, biopolymers such as xanthan gum and carboxymethyl cellulose (CMC) exhibit superior thermal and chemical stability. CMC, in particular, is advantageous due to its abundance, biodegradability, and strong thickening and emulsification capabilities, even in seawater. The combined application of surfactants and polymers known as SP flooding has demonstrated significant improvements in oil recovery, sometimes yielding up to 20% additional extraction of the OOIP [10, 11]. However, operational challenges such as emulsion stability, fluid compatibility, and scaling persist. Research has shown that integrating biopolymer-based CMC derived from agricultural waste like palm oil empty fruit bunches with surfactants such as sodium dodecylbenzene sulfonate (SDBS) can enhance rheological behavior and overall EOR performance [12, 13]. In parallel, interest in natural surfactants has grown due to their environmental safety, availability, and effective surface-active properties. Despite their potential, a comprehensive understanding of their mechanisms in EOR especially those not derived from microbial sources remains limited in the literature. Recent reviews have begun to explore their role in reducing IFT, altering wettability, and interacting with reservoir rocks, revealing their promise as sustainable alternatives to synthetic agents [14]. EOR methods particularly chemical strategies such as surfactant flooding are fundamentally aimed at improving microscopic and macroscopic sweep efficiencies by injecting energy or chemicals into reservoirs. These methods aim to disrupt existing oil/rock/brine interactions and mobilize capillary-held oil [14, 15].

Lignin, a widely abundant and cost-effective natural biopolymer, has recently gained attention as a sustainable candidate for enhanced oil recovery (EOR) applications. Despite its potential, the practical application of lignin in EOR has been limited due to its complex three-dimensional aromatic structure and inherently weak interfacial properties [2, 16]. Traditional modification techniques to improve lignin's surface activity often involve blending with other surfactants or processing under strong alkaline conditions. To overcome these challenges, recent advancements have introduced novel surfactants by integrating asymmetric Gemini structures into lignin molecules. These lignin-based Gemini surfactants (ALGS) demonstrate significantly improved salt resistance, emulsification, adsorption behavior, and wettability alteration capabilities, which are essential for EOR performance [17, 18]. Structurally, lignin is a heterogeneous macromolecule composed of phenylpropane units such as coniferyl (G), sinapyl (S), and p-coumaryl (H) alcohols—linked primarily through β-O-4 aryl ether bonds. The variability in its structure, influenced by plant source and isolation method, offers numerous functional groups such as phenolics, hydroxyls, and methoxyls for chemical modification [19]. These features contribute to lignin's excellent properties such as thermal stability, antibacterial activity, antioxidant capacity, and UV protection, enabling its transformation into high-value products like lignin nanoparticles (LNPs). LNPs possess amphiphilic surfaces, high surface charge, and strong adsorption energy at the oil-water interface, making them effective and environmentally friendly stabilizers for Pickering emulsions. Their ability to prevent droplet coalescence through rigid interfacial films not only enhances emulsion stability but also aligns with sustainability goals by repurposing lignin, a waste by-product of the pulp and paper industry, into functional EOR agents [19]. Furthermore, their use reduces $CO_2$ emissions and mitigates environmental pollution associated with lignin disposal. Complementing these developments, lignin-containing nanocellulose (LCN) has emerged as a promising material for EOR gel systems. By combining the thermal stability of lignin with the hydrophilicity of cellulose, LCN enables the formation of three-dimensional gel networks with superior rheological properties. These materials have demonstrated resilience under harsh reservoir conditions, including high temperature and salinity, and are being explored as alternatives to conventional crosslinkers like polyethyleneimine [20]. The thermal and interfacial properties of lignin have also been enhanced *via* a series of chemical modifications such as sulfonation, amination, and carboxylation. These processes improve lignin's compatibility with oil-wet surfaces and its solubility in aqueous phases, facilitating its incorporation into EOR surfactant systems [2]. For instance, sulfonated lignin-based zwitterionic surfactants synthesized from alkali lignin have demonstrated low interfacial tension (IFT) values, making them viable candidates for oil displacement applications [21]. In addition to their chemical versatility, LNPs exhibit high desorption-free energy, which translates to enhanced emulsion stability during oil recovery. However, studies indicate that their movement at the oil-water interface is relatively slow without external energy input, a factor that requires consideration in EOR system design [22]. Overall, lignin and its derivatives—as whether modified surfactants, nanoparticles, or nanocellulose hybrids—offer promising, sustainable alternatives for advancing EOR technologies. These materials combine environmental benefits with functional efficiency, addressing critical challenges in oil recovery such as high salinity, elevated temperatures, and the need for biodegradable, low-cost surfactants [23].

## 2. Renewable alternatives in petroleum applications

The petroleum industry is undergoing a transformative shift as global concerns about environmental sustainability, climate change, and resource depletion intensify. Traditionally reliant on fossil-based materials for extraction, processing, and production, the industry now faces mounting pressure to adopt greener and more sustainable alternatives. This transition is especially crucial in light of rising carbon emissions, environmental degradation, and the urgent need to achieve global climate targets, such as those outlined in the Paris Agreement. The shift from fossil-derived materials to bio-based alternatives is not only a response to environmental regulations but also a strategic move to ensure long-term operational resilience and resource security. Fossil fuels, though energy-dense and historically indispensable to industrial development, are finite and environmentally taxing. The extraction and utilization of petroleum contributes significantly to greenhouse gas (GHG) emissions, water contamination, soil degradation, and air pollution [24]. In contrast, bio-based materials derived from renewable biomass sources such as lignin, cellulose, starch, and plant oils—offer lower environmental footprints and are often biodegradable, non-toxic, and replenishable. Among these, *lignin*, the second most abundant natural polymer after cellulose, has emerged as a particularly promising resource for petroleum-related applications. It is a by-product of the pulp and paper industry and is typically underutilized or incinerated, resulting in both resource waste and environmental pollution [25].

The development of *lignin-based surfactants, nanoparticles, and gels* represents a crucial innovation for enhanced oil recovery (EOR) and other petroleum processes. By replacing synthetic, fossil-derived surfactants with renewable bio-based counterparts, the industry can significantly reduce its reliance on petrochemicals and contribute to the circular bioeconomy [26]. The environmental imperative driving this transition stems from multiple factors:

1. *Carbon emission reduction*: Bio-based materials, especially those from agricultural or forestry residues, are considered carbon-neutral or carbon-negative, depending on their life cycle. Replacing fossil-based surfactants with lignin-derived ones in EOR can reduce lifecycle emissions and help meet carbon reduction goals [24].

2. *Pollution mitigation*: Traditional petroleum additives and surfactants can persist in the environment, contaminating groundwater and affecting aquatic life. Bio-based materials like lignin, being biodegradable and non-toxic, reduce the ecological impact of oilfield chemicals [2].

3. *Waste valorization*: The use of lignin converts a major industrial waste into a valuable resource, contributing to waste minimization strategies. Around 50 million tons of lignin are generated annually worldwide, yet most of it is discarded or burned [21].

4. *Compliance with environmental regulations*: Regulatory frameworks across the globe are increasingly mandating the reduction of persistent and toxic chemicals in industrial applications. Bio-based chemicals can help oil and gas operators meet regulatory compliance more efficiently.

From an industrial perspective, the integration of renewable materials into petroleum operations offers several strategic advantages:

- *Cost competitiveness*: While initial R&D costs may be high, the abundance and low cost of biomass-derived feedstocks like lignin make them economically attractive in the long run [19].

- *Functionality enhancement*: Bio-based additives can be engineered to provide advanced functionalities. For example, lignin-derived Gemini surfactants have shown improved salt resistance, surface activity, and emulsification performance in EOR applications [17].

- *Performance in harsh conditions*: Lignin-containing nanocellulose (LCN) gels exhibit superior thermal stability and salinity resistance, making them suitable for high-temperature, high-salinity oil reservoirs [20].

- *Corporate sustainability goals*: Many oil companies have adopted Environmental, Social, and Governance (ESG) targets, and incorporating renewable alternatives supports broader decarbonization and sustainability objectives.

The petroleum industry's pivot toward bio-based materials such as lignin is both an environmental necessity and an industrial opportunity. These materials offer a viable pathway to reduce dependence on fossil-based chemicals, mitigate environmental harm, and align operations with global sustainability trends. With continued research and technological development, renewable biopolymers like lignin are poised to play a critical role in the future of petroleum applications.

## 3. Lignin processing

For industrial applications, lignin must be effectively isolated from lignocellulosic biomass using a variety of chemical and mechanical methods. These can broadly be categorized into two strategies: (i) processes that selectively solubilize hemicellulose and cellulose, leaving lignin as an insoluble residue, and (ii) processes in which lignin is dissolved or extracted, leaving polysaccharides intact [2]. The recovered lignin differs significantly in structure, molecular weight, and physicochemical properties depending on the biomass source and delignification method used [2]. The most prevalent industrial delignification techniques include kraft, sulfite, and organosolv pulping, while pyrolysis, steam explosion, alkaline hydrolysis, acidolysis, and enzymatic hydrolysis serve as less common alternatives. **Figure 1** summarizes industrial lignin extraction methods. Key parameters such as pH, temperature, chemical concentration, and pressure strongly affect extraction yield, solubility, and lignin's functional properties.

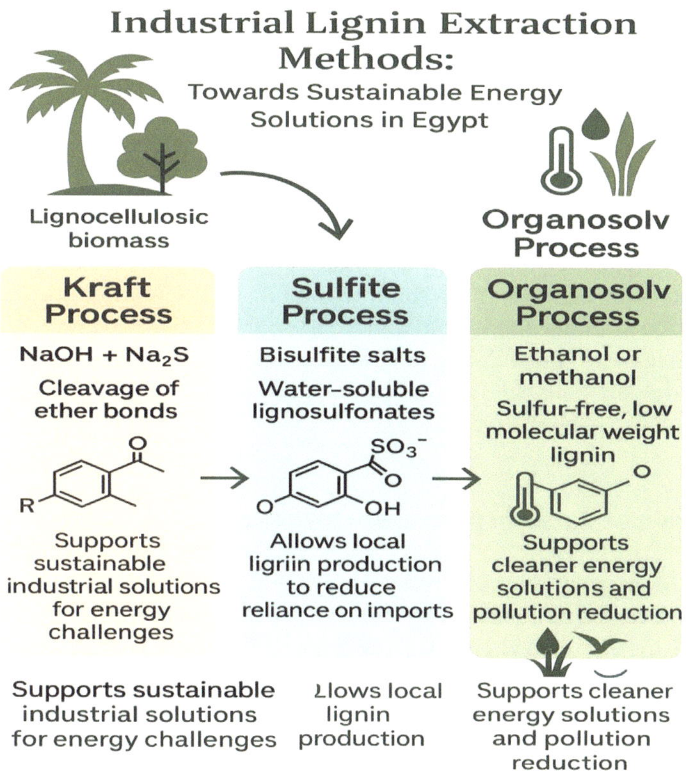


**Figure 1.**
*Industrial lignin extraction methods.*

### 3.1 Kraft pulping

The kraft process remains the dominant method in lignin extraction, generating approximately 75 million tons of lignin annually as a by-product of the pulp and paper industry. During this process, sodium hydroxide (NaOH) converts phenolic groups into quinomethide intermediates. These intermediates are then cleaved by hydrogen sulfide ($H_2S$) ions, primarily targeting α-carbon atoms in ether linkages, ultimately forming benzyl thiolate anions. Subsequent reactions, catalyzed by sodium sulfide ($Na_2S$), result in β-phenolate anions and regeneration of phenolic groups. Repetition of this cycle leads to extensive cleavage of ether linkages and generation of low-molecular-weight fragments. However, toward the end of the process, condensation reactions may occur, forming rigid C–C bonds and highly condensed lignin structures [2].

### 3.2 Sulfite pulping

In contrast, the sulfite pulping process yields approximately 1000 tons of lignosulfonate-based lignin annually. This method employs a heated aqueous solution of bisulfite and sulfite salts with variable counterions (e.g., $Ca^{2+}$, $Mg^{2+}$, $Na^+$, $NH_4^+$) across a broad pH range (1 to 13.5). The process centers on the sulfonation of aliphatic chains in lignin, forming water-soluble lignosulfonates that co-exist with hemicellulose in the pulping liquor. Separation of lignin from this aqueous phase typically involves techniques such as ultrafiltration, precipitation, or sugar fermentation followed by distillation [2].

### 3.3 Organosolv pulping

Organosolv pulping is a sulfur-free delignification method gaining traction for its environmental benefits and efficiency. It utilizes organic solvents—such as ethanol, methanol, acetic acid, or formic acid—with or without catalysts, under high temperatures (100–200°C) for short durations (30–150 minutes). This technique is particularly suitable for both woody and annual plants. Its sulfur-free nature avoids the emission of harmful gases associated with kraft and sulfite pulping and yields lignin with a lower molecular weight and higher purity. Furthermore, organosolv methods enable more controlled depolymerization, making the extracted lignin more amenable to downstream valorization in advanced applications such as bio-based surfactants, gels, and nanoparticles [2].

## 4. Application of lignin in petroleum processes

Lignin, a highly abundant and renewable biopolymer derived from lignocellulosic biomass, has emerged as a promising material for various applications in petroleum processes due to its unique chemical structure and functional versatility. Its phenolic and aromatic composition allows for chemical modifications that enhance surface activity, thermal stability, and salinity resistance key requirements in enhanced oil recovery (EOR) operations. Lignin-derived surfactants, such as lignin-based Gemini and zwitterionic types, have shown excellent potential in reducing the interfacial tension between oil and water, stabilizing emulsions, and altering reservoir wettability, which significantly improves oil displacement efficiency. Furthermore, lignin-containing nanocellulose (LCN) and lignin nanoparticles (LNPs) serve as stabilizers in Pickering emulsions and polymer gel systems used for profile control and conformance improvement in harsh reservoir conditions. These lignin-based materials offer an eco-friendly and cost-effective alternative to conventional petroleum-based chemicals, aligning with the industry's goals of sustainability and reduced environmental impact. In these sections, a short overview of different lignin applications will be discussed.

### 4.1 Application in enhanced oil recovery

Following primary and secondary recovery methods, a large portion of oil remains trapped in reservoir pores, necessitating enhanced oil recovery (EOR) techniques that modify the interfacial properties and improve fluid flow [27]. Among these, Pickering emulsions have emerged as effective agents for mobilizing residual oil by enhancing sweep efficiency and dislodging oil droplets through interfacial deformation. Recent advances spotlight lignin nanoparticles (LNPs) as sustainable, biodegradable stabilizers for such emulsions, offering superior thermal and salinity resistance, near-neutral wettability, and pH-responsive behavior that aids both oil displacement and separation [19]. For instance, LNP-based emulsions have shown stability at elevated temperatures and tunable destabilization, making them especially suitable for low-permeability reservoirs [3]. Beyond emulsification, lignin also serves as a feedstock for advanced surfactants and nanomaterials, such as the biphenyl sulfonate Gemini surfactant (BPSG), which outperforms conventional agents in interfacial activity and oil recovery. Moreover, lignin-derived nanomaterials enhance electrical EOR processes like microwave or electromagnetic heating by improving thermal conductivity and modifying rock wettability, all while reducing environmental risks [28]. As sustainability becomes a priority, lignin-based chemicals offer cost-effective and eco-friendly alternatives to synthetic counterparts, as supported by recent studies. However, challenges such as high-energy emulsification, complex oil-water separation, and limited field-scale data remain [29]. Future research should focus on scalable, energy-efficient, and stimuli-responsive lignin-based systems to enable broader application in petroleum operations.

#### *4.1.1 Challenges and future directions for lignin in enhanced oil recovery (EOR)*

The integration of lignin and its derivatives particularly lignin nanoparticles (LNPs) into enhanced oil recovery (EOR) processes represents a promising strategy toward sustainable petroleum technologies. However, several *technical, environmental, and economic challenges* must be addressed before large-scale industrial deployment becomes feasible.

##### *4.1.1.1 Technical limitations and performance constraints*

Despite their advantageous properties, lignin-based materials still face significant technical hurdles in EOR applications. One of the most persistent issues is the *inconsistency in lignin properties* due to variations in biomass source, extraction method, and chemical composition. These variations can significantly influence the surface charge, wettability, molecular weight, and stability of lignin derivatives, thereby affecting their emulsification performance and interaction with crude oil and reservoir rock [30]. In Pickering emulsion systems, *droplet stability under reservoir conditions* (e.g., high salinity, temperature, and pressure) is crucial. While LNPs show strong interfacial adsorption and thermal resistance, the *irreversible stability* of the resulting emulsions often complicates oil-water separation post-recovery [31]. Additionally, the *energy-intensive preparation methods*—such as high-pressure homogenization, ultrasonication, or microfluidization—pose significant scalability and cost-efficiency challenges [28, 31].

*4.1.1.2 Environmental and operational trade-offs*

While lignin is generally regarded as an eco-friendly material, *certain chemical modifications* used to enhance its surface activity such as sulfonation or grafting with hydrophobic chains—may introduce *ecotoxicity* or hinder biodegradability [3]. Furthermore, although lignin reduces reliance on synthetic surfactants and polymers, its performance under *harsh subsurface conditions* still lags behind some petroleum-based alternatives in terms of long-term interfacial stability and oil displacement efficiency [32]. Another significant operational concern is the *interaction of LNPs with reservoir rock*, which can lead to *pore plugging*, permeability reduction, or unpredictable retention behavior. This raises the need for an in-depth investigation into transport mechanisms and adsorption dynamics of lignin nanomaterials within porous media [28].

*4.1.1.3 Economic feasibility and commercial viability*

The economic competitiveness of lignin-based EOR agents is another barrier to adoption. Although lignin is abundantly available as a by-product of the pulp and paper industry, the *cost of refining, functionalizing, and processing* it into suitable forms (e.g., nanoparticles or surfactants) remain relatively high. Energy input during processing not only increases production costs but also *offsets the environmental benefits* if not managed sustainably [33]. The absence of *standardized industrial protocols* for LNP preparation and testing also hampers commercialization. Most studies remain at the lab or pilot scale, and there is a lack of *field-scale validation* of LNP-based EOR technologies, particularly under diverse geological formations [34].

*4.1.2 Future directions in EOR*

Despite these challenges, several research avenues can significantly improve the viability and impact of lignin-based EOR technologies:

*4.1.2.1 Development of smart, stimuli-responsive LNPs*

Creating *pH-, temperature-, or salinity-responsive* lignin-based emulsifiers can help address oil-water separation issues post-recovery. Such materials would allow emulsions to remain stable under reservoir conditions but break down in controlled surface environments, simplifying separation and minimizing waste [31].

*4.1.2.2 Eco-friendly functionalization techniques*

Advancements in *green chemistry* may lead to more sustainable ways to modify lignin for enhanced surface activity without compromising biodegradability. Enzymatic treatments and biocatalytic routes are particularly promising alternatives to conventional chemical grafting [35].

#### *4.1.2.3 Hybrid systems with other bio-based or mineral additives*

Lignin can be synergistically combined with *silica nanoparticles, clay minerals*, or *cellulose derivatives* to form *hybrid emulsifiers or nanofluids* with improved interfacial properties, viscosity control, and reservoir compatibility [17].

#### *4.1.2.4 Standardization and scale-up*

Future work should focus on the *standardization of LNP production protocols*, field-testing under various geological settings, and *life cycle assessments* (LCA) to establish their commercial viability compared to synthetic chemicals. Public-private partnerships and pilot projects in collaboration with oil companies can facilitate this transition.

#### *4.1.2.5 Computational modeling and pore-scale simulations*

Incorporating *computational fluid dynamics (CFD)* and *molecular dynamics (MD) simulations* will enhance understanding of LNP behavior at the oil-water-rock interface, optimize formulation, and reduce trial-and-error in product development [28]. Lignin and its derivatives present an exciting frontier in environmentally responsible EOR. However, for these bio-based materials to transition from academic research to industrial implementation, critical challenges in processing, consistency, field performance, and cost must be overcome. By integrating green chemistry, responsive materials, hybrid technologies, and interdisciplinary modeling, the petroleum industry can unlock the full potential of lignin for sustainable oil recovery.

### 4.2 Reduction of interfacial tension (IFT) in CEOR using lignin-based surfactants

In chemical enhanced oil recovery (CEOR), one of the most critical strategies for mobilizing trapped crude oil is the reduction of interfacial tension (IFT) between oil and water phases. Traditional surfactants classified as anionic, nonionic, cationic, or amphoteric have long been utilized for this purpose, as they can significantly reduce IFT and improve oil displacement efficiency. Additionally, surfactants often aid in modifying rock wettability and enhancing the mobility ratio, further boosting recovery rates. However, despite their wide industrial usage, many synthetic surfactants present considerable drawbacks at the field scale, particularly related to *cost, environmental sustainability, and safety* [36]. With increasing scrutiny over the environmental impacts of synthetic chemicals, there has been a marked shift toward *bio-based surfactants*, particularly those derived from lignocellulosic biomass. *Lignin*, an abundant and renewable byproduct of the paper and bioethanol industries, has emerged as a promising candidate for developing eco-friendly surfactants. Its aromatic-rich, amphiphilic structure makes it suitable for chemical modification into surface-active agents capable of reducing IFT and interacting at the oil-water interface [37]. Recent studies have shown that lignin-derived surfactants such as lignosulfonates, lignin betaines, and Gemini-type surfactants synthesized from modified lignin—can achieve substantial IFT reduction while exhibiting *lower ecotoxicity and higher biodegradability* than traditional synthetic surfactants. For instance, Chen et al. [17]

reported the successful synthesis of a biphenyl sulfonate Gemini surfactant (BPSG) from lignin, which displayed effective IFT reduction and favorable surface activity compared to conventional agents like sodium dodecylbenzene sulfonate (SDBS). In addition to being environmentally preferable, *lignin-based surfactants are often more cost-effective*, especially when produced from industrial waste streams such as black liquor from kraft pulping. This aligns with circular economy principles, where waste is valorized into value-added products for industrial applications. Despite these advantages, challenges remain. Lignin's *inherent structural heterogeneity*—influenced by plant source, extraction method, and processing—can lead to variability in surfactant performance. Moreover, scaling up the production of tailored lignin-based surfactants requires robust process optimization and consistent quality control. The *efficacy of lignin surfactants under reservoir-relevant conditions* (e.g., high temperature, salinity, and pressure) also needs further investigation [28].

### 4.3 Lignin nanoparticles for foam stabilization

Aqueous foam systems comprising gas bubbles dispersed in a liquid medium play a vital role in several industrial domains such as detergents, pharmaceuticals, food processing, mineral flotation, and particularly, enhanced oil recovery (EOR). In EOR, foams are used to improve the sweep efficiency of gases (e.g., $CO_2$ or steam) by mitigating gas channeling and improving fluid mobility control. However, one of the biggest technical obstacles is *foam instability*, especially under harsh conditions such as high temperatures, high salinity, and mechanical shear, which are common in oil reservoirs. Traditionally, *surfactants* have been employed to generate and stabilize foam by reducing interfacial tension (IFT) and delaying the coalescence of gas bubbles. Nonetheless, these surfactant-only foams are prone to *rapid liquid drainage*, leading to the thinning and eventual collapse of foam films—particularly under extreme thermal or mechanical stress [38]. To counteract this, *co-stabilizers* such as water-soluble polymers and *worm-like micelles* have been introduced to increase viscosity and reinforce foam films. In recent years, *nanoparticles* such as $SiO_2$, $Al_2O_3$, and $TiO_2$ have been employed as stabilizing agents to physically strengthen foam lamellae, thereby extending foam lifetime and mechanical integrity [39]. However, these synthetic additives can be costly, difficult to degrade, and environmentally concerning. This has led to increased interest in *bio-based stabilizers such as lignin nanoparticles (LNPs)*. Lignin, a polyphenolic biopolymer abundant in plant biomass, possesses intrinsic properties such as high *thermal stability, amphiphilicity*, and *functional groups* (hydroxyl, methoxy) that can facilitate foam stabilization. Research by Fan et al. [40] demonstrated that when LNPs were combined with *alpha-olefin sulfonate (AOS)* a thermally stable anionic surfactant, the resulting composite foam showed significantly *improved static and dynamic stability*. At temperatures up to 423 K, the foam half-life was increased tenfold compared to surfactant-only systems, with enhanced resistance to shear and structural collapse. These improvements were attributed to the *synergistic interactions* between LNPs and surfactants both in the bulk solution and at the gas–liquid interface. LNPs facilitated *tighter surfactant packing*, created *thicker and more cohesive foam films*, and reduced surface tension more effectively. These structural enhancements slowed gas diffusion, improved liquid retention, and increased the mechanical robustness of the foam [40]. Moreover, compared to inorganic nanoparticles or synthetic polymers, LNPs offer several *sustainability benefits*: They are *biodegradable*, *non-toxic*, and *derived from renewable resources*, making them highly suitable for environmentally responsible EOR operations and broader applications in consumer goods.

### 4.4 Electrolytic water splitting reactions

Recent advancements in renewable electrochemical technologies have increasingly focused on integrating lignin with nanomaterials to develop high-performance, sustainable catalysts for energy conversion. One such innovation is the development of chemically modified electrodes composed of carbon nanotubes (CNTs), lignin, and polysaccharides. These bio-nanocomposite electrodes show considerable promise in a range of electrochemical reactor (ECR) applications, particularly in electrolytic water splitting reactions, where hydrogen and oxygen gases are produced *via* the oxygen evolution reaction (OER) and hydrogen evolution reaction (HER), respectively [41]. Lignin, as a natural aromatic polymer, contributes to these composites not only due to its renewability and abundance but also due to its redox-active phenolic and methoxy groups. When combined with CNTs that offer excellent electrical conductivity, large surface area, and structural support—lignin can improve charge transfer kinetics and stabilize catalytic activity. These hybrid CNT-lignin materials provide a cost-effective alternative to noble metal catalysts (such as Pt, Ir, or Ru), whose limited availability and high cost have hindered large-scale deployment of traditional water-splitting technologies [42]. In particular, the incorporation of lignin and biopolymers such as cellulose or chitosan into CNT-based electrode systems has shown synergistic effects: enhancing catalytic efficiency, improving stability under operational conditions, and enabling tunability of the catalyst's physicochemical properties. Moreover, the use of lignin in these systems supports the broader goal of circular bioeconomy by valorizing lignocellulosic waste streams [43]. Electrochemical studies have demonstrated that these CNT-lignin-biopolymer composites can effectively catalyze both OER and HER under alkaline conditions, with improved durability and minimal degradation over multiple cycles. Additionally, their application extends beyond water splitting to include fuel oxidation, oxygen reduction reactions (ORR), $CO_2$ reduction, and proton exchange membrane fuel cells, establishing a broad platform for clean energy generation.

### 4.5 Marine fouling and lignin-based antifouling coatings

Marine biofouling the accumulation of microorganisms, algae, and small animals on submerged surfaces—is a persistent and costly issue for marine equipment and offshore infrastructure. It accelerates corrosion, increases hydrodynamic drag, raises fuel consumption, and reduces the service life of vessels and underwater installations. Traditional antifouling solutions rely heavily on petroleum-derived polymers such as silicone, polyurethane, and epoxy resins. These coatings often contain toxic biocidal agents, which, while initially effective, leach harmful substances into marine ecosystems over time and contribute to the development of biocide-resistant fouling organisms [44]. In response to environmental regulations and rising ecological awareness, recent research has shifted toward developing sustainable antifouling coatings using naturally derived materials. A promising candidate in this context is *lignin*, a polyphenolic biopolymer abundantly found in plant cell walls. Lignin is biodegradable, non-toxic, and offers unique antimicrobial and antioxidant properties, making it a suitable base material for eco-friendly coatings [45]. A study by Wu et al. [46] introduced a novel green antifouling and wear-resistant coating formulated with a hydrophilic-to-hydrophobic ratio of 2:8, incorporating lignin to enhance surface performance. The lignin-based coating increased the water contact angle by 14.5%, significantly improving surface hydrophobicity. This property is critical in antifouling performance as it reduces the adhesion of marine organisms to the surface. Moreover, lignin's molecular structure, rich in functional hydroxyl and methoxy groups, allows strong interactions with other coating constituents, improving mechanical durability and resistance to degradation under marine conditions. Beyond its functional role, lignin contributes to sustainability goals. By replacing petroleum-based polymers, lignin reduces the environmental footprint of coating production and application. It also minimizes the release of toxic substances into the aquatic environment, offering a safer alternative to conventional biocidal antifouling strategies. Importantly, lignin's inherent antimicrobial activity provides passive protection, without the need for added chemical biocides, thereby reducing the risk of resistance development among marine fouling organisms [47]. Despite these benefits, challenges remain in optimizing the adhesion, long-term stability, and large-scale application of lignin-based coatings in harsh marine environments. Future research should focus on enhancing lignin dispersion within coating matrices, investigating hybrid formulations (e.g., lignin with silica or nano-clay), and conducting long-duration field trials in diverse marine settings. Moreover, lifecycle assessments are needed to confirm the full environmental advantages of lignin over conventional materials.

### 4.6 Advanced water solutions: Lignin-based hydrogels for wastewater treatment

Biomass-derived polymers are increasingly recognized as sustainable alternatives to petroleum-based plastics in wastewater treatment, owing to their biodegradability, biocompatibility, cost-effectiveness, and environmental safety. Among these, lignin second only to cellulose in abundance is the primary renewable polyaromatic resource on Earth and serves as a critical feedstock for the synthesis of high-performance materials like hydrogels [48, 49]. Structurally, lignin is a three-dimensional, amorphous, and crosslinked polymer composed of phenylpropanoid units such as coniferyl, sinapyl, and p-coumaryl alcohols. These units give rise to functional groups including hydroxyl, methoxyl, and carbonyl moieties, which contribute to lignin's hydrophilicity and reactive potential—ideal traits for hydrogel formation [50]. Lignin can be extracted through several processes, including the Kraft, soda, and sulfite pulping methods, with the Kraft process being the most common industrial technique. Additionally, green solvents like ionic liquids have been explored for more sustainable lignin recovery [51]. The recovered lignin can be used in its native form or chemically modified to produce lignin-based hydrogels (LBHs), a class of materials exhibiting high water absorption, mechanical strength, and tailored functionality. These hydrogels are prepared *via* covalent crosslinking of water-soluble monomers or modified lignin derivatives to form a stable, three-dimensional polymer network [48, 52]. Lignin-based hydrogels have attracted growing interest due to their eco-friendly profile and applicability across various domains. In wastewater treatment, LBHs demonstrate strong adsorption capacity for heavy metals, dyes, and other organic pollutants, making them viable substitutes for synthetic absorbents [53]. Their structural rigidity—conferred by aromatic rings enhances durability and mechanical performance, while the hydroxyl groups enable hydrogen bonding, swelling behavior, and pollutant capture. These properties also make LBHs suitable for applications in agriculture (as water-retentive soil additives), oil recovery (as fluid-loss control agents), and biomedicine (for drug delivery and wound care) [54, 55]. Despite their promising attributes, LBHs face several development challenges. These include variability in lignin composition based on the extraction method and feedstock, difficulties in reproducible crosslinking, and the need for scalable, low-cost synthesis techniques. Moreover, their long-term stability in harsh wastewater conditions and regeneration capacity requires further investigation. Nonetheless, advances in functionalization methods—such as grafting, copolymerization, and incorporation of nanomaterials are addressing these issues and expanding the performance envelope of LBHs [56].

### 4.7 Oil-water separation films

Oil-water separation is a critical challenge in environmental cleanup and industrial wastewater treatment, particularly in applications like marine oil spill response and oily effluent remediation. Traditional polymeric membranes and absorbent materials often fall short due to poor durability under harsh environmental conditions, limited recyclability, or environmental impact from their production. In this context, integrating bio-based materials such as lignin into separation membranes offers a green, sustainable alternative. Kassaun and Fatehi [57] investigated the development of a high-performance oil-water separation film by incorporating a lignin-based copolymer into natural rubber (NR). Natural rubber is commonly used for its flexibility and biocompatibility but suffers from degradation when exposed to heat and moisture. The study addressed these weaknesses by embedding a chemically modified lignin copolymer—produced *via* water-based copolymerization of kraft lignin with aminopropyl/fluoro-silsesquioxane—into the NR matrix. The resulting composite, referred to as W20 (containing 20% of the lignin-silsesquioxane copolymer), demonstrated an 81% improvement in mechanical strength compared to NR composites with unmodified kraft lignin alone. Notably, W20 retained superior performance under thermal and moisture stress, offering enhanced durability and stability [57]. The W20 film also exhibited *superhydrophobic properties*, maintaining water repellency even after mechanical abrasion, and achieved *96% efficiency in oil absorption and separation*, indicating strong potential for use in environmental cleanup operations. These enhancements are attributed to the structural benefits of the lignin-based hybrid material, which provides both thermal stability and interfacial modification capabilities. Lignin's sustainability, biodegradability, and antioxidant properties make it a highly attractive candidate for green material design [58]. However, its inherent polarity and tendency to aggregate within hydrophobic matrices like NR can reduce composite performance. Various strategies including lignin nanoparticle formation, hybridization with siloxanes, and chemical grafting have been proposed to improve the compatibility with NR and enhance mechanical properties [59]. In W20, the fluorinated silsesquioxane modification effectively overcame the typical issues of phase separation and aggregation, resulting in a uniformly dispersed and mechanically robust material. Despite these advancements, challenges remain. Many modification routes still rely on environmentally hazardous solvents or high-energy inputs. Moreover, excessive lignin loading even in modified form can lead to reduced elasticity and compromised film performance. Therefore, continued innovation in eco-friendly lignin derivatives and scalable green synthesis methods is crucial for advancing lignin-based oil-water separation technologies. Lignin-modified NR films like W20 provide a promising platform for developing sustainable, high-performance oil-water separation systems. These materials not only reduce environmental impact through bio-based sourcing but also demonstrate technical viability in harsh operational environments, pushing the boundaries of green engineering applications.

## 5. Challenges of lignin application in the petroleum industry

### 5.1 Unclear stabilization mechanisms

The stabilization mechanisms of Pickering emulsions by lignin nanoparticles (LNPs) remain insufficiently understood, largely due to the inherently heterogeneous and irregular nature of lignin's polymeric structure. Composed of randomly polymerized phenylpropanoid units, lignin forms a complex, three-dimensional network that poses significant challenges for theoretical modeling and predictive analysis. To advance the effective application of LNPs in enhanced oil recovery, a deeper molecular-level understanding is needed particularly regarding how these nanoparticles adsorb at oil-water interfaces, reduce interfacial tension, and maintain emulsion stability under the variable and dynamic conditions typical of reservoir environments.

### 5.2 Lack of scalable production methods

Current methods for synthesizing lignin nanoparticles (LNPs) such as solvent exchange, anti-solvent precipitation, and ultrasonication are typically limited to laboratory-scale settings, require high energy input, and produce only small quantities. The lack of cost-effective, continuous, and high-throughput production techniques remains a major barrier to the widespread adoption of LNPs in oilfield applications. Addressing this challenge will require the development of scalable, environmentally friendly, and economically viable manufacturing processes to enable industrial deployment.

### 5.3 Biological toxicity and environmental safety

Although lignin nanoparticles (LNPs) are derived from natural and renewable sources, their nanoscale size and high surface reactivity raise concerns about potential bioaccumulation, ecotoxicity, and unintended biological interactions—particularly in applications involving prolonged environmental exposure, such as oil spill remediation. To ensure safe and responsible use, comprehensive studies on their toxicity, environmental fate, and long-term behavior are essential. Additionally, strategies such as surface modification or incorporation of biocompatible stabilizers may be necessary to mitigate these risks and enhance their environmental safety profile.

### 5.4 Limited thermal resistance

Lignin nanoparticles (LNPs), similar to other biomass-derived materials, exhibit limited thermal stability, which restricts their use in high-temperature petroleum applications such as steam-assisted gravity drainage (SAGD) or high-pressure, high-temperature (HPHT) reservoirs. To overcome this limitation, research should focus on enhancing their heat resistance and mechanical robustness through chemical modifications such as cross-linking or grafting or by developing hybrid systems like lignin-silica composites that can withstand the harsh thermal conditions encountered in advanced oil recovery processes.

## 6. Future prospects and research directions

### 6.1 Optimization of LNP structures by lignin source

- Different types of lignin (e.g., kraft, organosolv, soda) yield LNPs with *distinct morphologies, surface charges, and functional group profiles*.

- Further *comparative studies* are needed to determine which lignin types and processing methods offer the best performance for oil recovery, emulsion stability, and interfacial activity.

### 6.2 Field-scale validation

- Most studies remain *limited to lab-scale testing*. There is an urgent need for *pilot-scale and real-field trials* to evaluate the effectiveness of LNPs in actual reservoir conditions, including:

  o Conventional chemical flooding (e.g., surfactant-polymer-LNP systems)

  o Electrically enhanced oil recovery (EEOR)

  o High-salinity and high-temperature reservoir environments

### 6.3 Development of magnetic LNPs

- *Magnetic lignin-based nanoparticles (M-LNPs)* represent a highly promising subfield, combining magnetic retrievability with environmental safety.

- M-LNPs offer opportunities for *stimulus-responsive oil recovery*, targeted delivery, and *post-treatment separation*, but more research is needed to fine-tune their synthesis and assess long-term stability and recyclability.

### 6.4 Integration with other green technologies

- LNPs could be integrated into *hybrid green oilfield systems*, including:

  o Lignin-based surfactants

  o Biopolymer gels

  o Nanocellulose-lignin composites

- These combinations may create *synergistic effects*, improving oil mobility control, reservoir sweep efficiency, and sustainability.

Lignin-based nanoparticles present a versatile and sustainable platform for various petroleum sector applications, from oil spill remediation to enhanced oil recovery. However, their practical deployment requires overcoming key technical and environmental challenges, particularly regarding *scalability, thermal stability, and toxicity*.

With targeted research and industrial collaboration, LNPs could become an essential tool in the oil industry's shift toward cleaner, more sustainable technologies.

## Acknowledgements

The author clarify that *ChatGPT was used for language refinement only*, not for generating original content.

## Author details

Abdelaziz El-hoshoudy[1,2]

1 PVT Lab, Production Department, Egyptian Petroleum Research Institute, Cairo, Egypt

2 PVT Service Center, Egyptian Petroleum Research Institute, Cairo, Egypt

*Address all correspondence to: azizchemist@yahoo.com
and abdelaziz.nasr@epri.sci.eg

## References

[1] El-hoshoudy A, Salem KG. Investigation of sulfonated polyethylene glycol-polyacrylate copolymer for enhanced oil recovery through core and field-scale simulation. Journal of Molecular Liquids. 2025;**420**:126820

[2] Shafiq A, Bhatti IA, Amjed N, Zeshan M, Zaheer A, Kamal A, et al. Lignin derived polyurethanes: Current advances and future prospects in synthesis and applications. European Polymer Journal. 2024:112899

[3] Chen S, Zhou X, Yang J, Dai Y, Wang W, Jiang W, et al. Study on synthesis and properties of novel bisphenyl sulphonate gemini surfactant based on lignin for enhanced oil recovery. Journal of Molecular Liquids. 2023;**390**:123072

[4] El-hoshoudy A, Mansour E, Gaffer AK, Atta AM. Novel green strategy based on combination of $CO_2$/protic poly (ionic liquids) for enhancing crude oil recovery. Energy & Fuels. 2024;**39**(1):538-558

[5] Abd-Elaal AA, Tawfik SM, Abd-Elhamid A, Salem KG, El-Hoshoudy A. Experimental and theoretical investigation of cationic-based fluorescent-tagged polyacrylate copolymers for improving oil recovery. Scientific Reports. 2024;**14**(1):27689

[6] Salem KG, Salem A, Gomaa S, El-hoshoudy A. Nanoparticles-Assisted Polymer Flooding: A New Approach toward Enhanced Oil Recovery. London, UK: IntechOpen; 2025

[7] Rahman AFA, Arsad A, Sidek A, Abdurrahman M. Wettability alteration and interfacial tension reduction on high salinity of anionic/nonionic surfactant. Materials Today: Proceedings. 2024;**110**:82-86

[8] Salem KG, Salem AM, Tantawy MA, Gawish AA, Gomaa S, El-hoshoudy A. A comprehensive investigation of nanocomposite polymer flooding at reservoir conditions: New insights into enhanced oil recovery. Journal of Polymers and the Environment. 2024;**32**(11):5915-5935

[9] El-hoshoudy A. Assessment of biopolymers rheological properties and flooding results under reservoir conditions for enhanced oil recovery processes. Egyptian Journal of Chemistry. 2024;**67**(9):231-244

[10] Abou-alfitooh SA, El-Hosiny F, El-Hoshoudy A. Experimental and computational study of modified biopolymer xanthan gum with synthetic vinyl monomers for enhanced oil recovery. Journal of Polymers and the Environment. 2024;**32**:1-20

[11] Salem KG, Tantawy MA, Gawish AA, Salem AM, Gomaa S, El-hoshoudy A. Key aspects of polymeric nanofluids as a new enhanced oil recovery approach: A comprehensive review. Fuel. 2024;**368**:131515

[12] Yusuf S, Junin R, Sidek MM, Agi A, Fuad MA, Rosli N, et al. Screening the synergy of sodium dodecylbenzenesulfonate and carboxymethyl cellulose for surfactant-polymer flooding. Petroleum Research. 2022;**7**(3):308-317. DOI: https://doi org/101016/J PTLRS 2021;2

[13] Khattab H, Gawish AA, Hamdy A, Gomaa S, El-Hoshoudy A. Assessment of a novel xanthan gum-based composite for oil recovery improvement at reservoir conditions; assisted with simulation and economic studies. Journal of Polymers and the Environment. 2024;**32**(7):3363-3391

[14] Atta DY, Negash BM, Yekeen N, Habte AD. A state-of-the-art review on the application of natural surfactants in enhanced oil recovery. Journal of Molecular Liquids. 2021;**321**:114888

[15] Khattab H, Gawish AA, Gomaa S, Hamdy A, El-Hoshoudy A. Assessment of modified chitosan composite in acidic reservoirs through pilot and field-scale simulation studies. Scientific Reports. 2024;**14**(1):10634

[16] Salem KG, Tantawy MA, Gawish AA, Gomaa S, El-Hoshoudy A. Nanoparticles assisted polymer flooding: Comprehensive assessment and empirical correlation. Geoenergy Science and Engineering. 2023;**226**:211753

[17] Chen S, Lei Q, Cai G, Wang F, Hu J, Li X, et al. Synthesis, properties and application of novel asymmetric Gemini surfactants based on alkali lignin. Journal of Molecular Liquids. 2025;**426**:127296

[18] El-Hoshoudy AN. Experimental and theoretical investigation of glycol-based hydrogels through waterflooding processes in oil reservoirs using molecular dynamics and dissipative particle dynamics simulation. ACS Omega. 2021;**6**(45):30224-30240

[19] Li M-C, Zhang Y, Sun J, Lv K, Huang X, Meng X, et al. Lignin nanoparticle-stabilized Pickering emulsion: Mechanism, influencing parameter, and emerging application. Advances in Colloid and Interface Science. 2025:103476

[20] Yang Z-G, Gao F, Guo J-X, Gao Q-Y, Xu Y-Y, Yang H, et al. Improving temperature and salinity tolerance of polymer gels by modifying with lignin-containing nanocellulose. Fuel. 2025;**384**:134019

[21] Chen S, Liu H, Sun H, Yan X, Wang G, Zhou Y, et al. Synthesis and physiochemical performance evaluation of novel sulphobetaine zwitterionic surfactants from lignin for enhanced oil recovery. Journal of Molecular Liquids. 2018;**249**:73-82

[22] Wang D, Zhu Y-L, Zhao Y, Li CY, Mukhopadhyay A, Sun Z-Y, et al. Brownian diffusion of individual janus nanoparticles at water/oil interfaces. ACS Nano. 2020;**14**(8):10095-10103

[23] Bai Y, Pu W, Jin X, Shen C, Ren H. Review of the micro and macro mechanisms of gel-based plugging agents for enhancing oil recovery of unconventional water flooding oil reservoirs. Journal of Molecular Liquids. 2024:124318

[24] Shah SWA, Xu Q, Ullah MW, Sethupathy S, Morales GM, Sun J, et al. Lignin-based additive materials: A review of current status, challenges, and future perspectives. Additive Manufacturing. 2023;**74**:103711

[25] Roy A, Fajardie P, Lepoittevin B, Baudoux J, Lapinte V, Caillol S, et al. CNSL, a promising building blocks for sustainable molecular design of surfactants: A critical review. Molecules. 2022;**27**(4):1443

[26] Wu Q, Liu T, Xu X, Yang J. A lignin-based Zwitterionic surfactant facilitates heavy oil viscosity reduction via interfacial modification and molecular aggregation disruption in high-salinity reservoirs. Molecules. 2025;**30**(11):2419

[27] Niro CM, Medeiros JA, Freitas JA, Azeredo HM. Advantages and challenges of Pickering emulsions applied to bio-based films: A mini-review. Journal of the Science of Food and Agriculture. 2021;**101**(9):3535-3540

[28] Juyal A, Bhadauriya DS, Sharma R, Deka B. Recovery of heavy crude oil with electrical enhanced oil recovery using lignin nanoparticles: A review. Materials Today: Proceedings. 2023;**99**:154-161

[29] El-Masry JF, Maalouf E, Abbas AH, Bou-Hamdan KF. Advancements in green materials for chemical enhanced oil recovery: A review. Petroleum. 2025;**11**:259-276

[30] Gosselink RJA. Lignin as a Renewable Aromatic Resource for the Chemical Industry. Ann Arbor, Michigan: Wageningen University and Research; 2011

[31] Gao K, Liu J, Li X, Gojzewski H, Sui X, Vancso GJ. Lignin nanoparticles as highly efficient, recyclable emulsifiers for enhanced oil recovery. ACS Sustainable Chemistry & Engineering. 2022;**10**(29):9334-9344

[32] Harry D, Horton D, Durham D, Constable DJ, Gaffney S, Moore J, et al. Grand challenges and opportunities for greener chemical alternatives in hydraulic fracturing: A perspective from the ACS green chemistry institute oilfield chemistry roundtable. Energy & Fuels. 2020;**34**(7):7837-7846

[33] Yadav P, Athanassiadis D, Antonopoulou I, Rova U, Christakopoulos P, Tysklind M, et al. Environmental impact and cost assessment of a novel lignin production method. Journal of Cleaner Production. 2021;**279**:123515

[34] Lee JG, Larive LL, Valsaraj KT, Bharti B. Binding of lignin nanoparticles at oil–water interfaces: An ecofriendly alternative to oil spill recovery. ACS Applied Materials & Interfaces. 2018;**10**(49):43282-43289

[35] Zhu W, Theliander H. Precipitation of lignin from softwood black liquor: An investigation of the equilibrium and molecular properties of lignin. BioResources. 2015;**10**(1):1696-1714

[36] Nowrouzi I, Mohammadi AH, Manshad AK. Water-oil interfacial tension (IFT) reduction and wettability alteration in surfactant flooding process using extracted saponin from Anabasis Setifera plant. Journal of Petroleum Science and Engineering. 2020;**189**:106901

[37] Pinheiro Pires AP, Arauzo J, Fonts I, Domine ME, Fernandez Arroyo A, Garcia-Perez ME, et al. Challenges and opportunities for bio-oil refining: A review. Energy & Fuels. 2019;**33**(6):4683-4720

[38] Wang Y, Zhang Y, Liu Y, Zhang L, Ren S, Lu J, et al. The stability study of $CO_2$ foams at high pressure and high temperature. Journal of Petroleum Science and Engineering. 2017;**154**:234-243

[39] Bhatt S, Saraf S, Bera A. Perspectives of foam generation techniques and future directions of nanoparticle-stabilized CO2 foam for enhanced oil recovery. Energy & Fuels. 2023;**37**(3):1472-1494

[40] Fan X, Guan X, Zhang M, Liu Y, Li Y. Aqueous foam synergistically stabilized by the composite of lignin nanoparticles and surfactant. Colloids and Surfaces A: Physicochemical and Engineering Aspects. 2022;**643**:128727

[41] Khorsandi Z, Nasrollahzadeh M, Kruppke B, Abbasi A, Khonakdar HA. Research progress in the preparation and application of lignin-and polysaccharide-carbon nanotubes for renewable energy conversion reactions. Chemical Engineering Journal. 2024:150725

[42] Colmenares JC, Varma RS, Nair V. Selective photocatalysis of lignin-inspired chemicals by integrating

hybrid nanocatalysis in microfluidic reactors. Chemical Society Reviews. 2017;**46**(22):6675-6686

[43] Tortolini C, Capecchi E, Tasca F, Pofi R, Venneri MA, Saladino R, et al. Novel nanoarchitectures based on lignin nanoparticles for electrochemical eco-friendly biosensing development. Nanomaterials. 2021;**11**(3):718

[44] Callow JA, Callow ME. Trends in the development of environmentally friendly fouling-resistant marine coatings. Nature Communications. 2011;**2**(1):244

[45] Beisl S, Miltner A, Friedl A. Lignin from micro-to nanosize: Production methods. International Journal of Molecular Sciences. 2017;**18**(6):1244

[46] Wu Z, Thoresen PP, Maršík D, Matsakas L, Kulišová M, Fous K, et al. High acetone soluble organosolv lignin extraction and its application towards green antifouling and wear-resistant coating. International Journal of Biological Macromolecules. 2024;**282**:137456

[47] Ghaffar SH, Fan M. Lignin in straw and its applications as an adhesive. International Journal of Adhesion and Adhesives. 2014;**48**:92-101

[48] Raj K, Vora T, PadmaPriya G, Lal B, Devi A, Sharma R, et al. A comprehensive review of sustainable hydrogels from lignin for advanced wastewater solutions. International Journal of Biological Macromolecules. 2025:139963

[49] Thakur VK, Thakur MK, Raghavan P, Kessler MR. Progress in green polymer composites from lignin for multifunctional applications: A review. ACS Sustainable Chemistry & Engineering. 2014;**2**(5):1072-1092

[50] Kai D, Tan MJ, Chee PL, Chua YK, Yap YL, Loh XJ. Towards lignin-based functional materials in a sustainable world. Green Chemistry. 2016;**18**(5):1175-1200

[51] El Mansouri N-E, Salvadó J. Structural characterization of technical lignins for the production of adhesives: Application to lignosulfonate, Kraft, soda-anthraquinone, organosolv and ethanol process lignins. Industrial Crops and Products. 2006;**24**(1):8-16

[52] Mondal AK, Uddin MT, Sujan S, Tang Z, Alemu D, Begum HA, et al. Preparation of lignin-based hydrogels, their properties and applications. International Journal of Biological Macromolecules. 2023;**245**:125580

[53] Naseer A, Jamshaid A, Hamid A, Muhammad N, Ghauri M, Iqbal J, et al. Lignin and lignin based materials for the removal of heavy metals from waste water-an overview. Zeitschrift für Physikalische Chemie. 2019;**233**(3):315-345

[54] Ding W, Sun H, Li X, Li Y, Jia H, Luo Y, et al. Environmental applications of lignin-based hydrogels for Cu remediation in water and soil: Adsorption mechanisms and passivation effects. Environmental Research. 2024;**250**:118442

[55] Kaur R, Sharma R, Chahal GK. Synthesis of lignin-based hydrogels and their applications in agriculture: A review. Chemical Papers. 2021;**75**(9):4465-4478

[56] Sethupathy S, Morales GM, Gao L, Wang H, Yang B, Jiang J, et al. Lignin valorization: Status, challenges and opportunities. Bioresource Technology. 2022;**347**:126696

[57] Kassaun BB, Fatehi P. Superhydrophobic lignin reinforced

rubber film as oil water separator. Sustainable. Materials and Technologies. 2025;**45**:e01444

[58] Figueiredo P, Lintinen K, Hirvonen JT, Kostiainen MA, Santos HA. Properties and chemical modifications of lignin: Towards lignin-based nanomaterials for biomedical applications. Progress in Materials Science. 2018;**93**:233-269

[59] Laurichesse S, Avérous L. Chemical modification of lignins: Towards biobased polymers. Progress in Polymer Science. 2014;**39**(7):1266-1290